Sustainable Biorefineries

Topical Conference at the 2008 AIChE Annual Meeting

Philadelphia, Pennsylvania
16 - 21 November 2008

ISBN: 978-1-61567-201-1

Printed from e-media with permission by:

Curran Associates, Inc.
57 Morehouse Lane
Red Hook, NY 12571

CURRAN ASSOCIATES INC. proceedings.com

Some format issues inherent in the e-media version may also appear in this print version.

Printed by Curran Associates, Inc. (2009)

For permission requests, please contact AIChE
at the address below.

AIChE
3 Park Avenue
New York, NY 10016-5991

Phone: (203) 702-7660
Fax: (203) 775-5177

www.aiche.org

Additional copies of this publication are available from:

Curran Associates, Inc.
57 Morehouse Lane
Red Hook, NY 12571 USA
Phone: 845-758-0400
Fax: 845-758-2634
Email: curran@proceedings.com
Web: www.proceedings.com

TABLE OF CONTENTS

"Bioenergy Conversion and Science – Challenges in Making Cellulosic Fuels"

Brian H. Davison
BioEnergy Science Center, and
Chief Scientist, Systems Biology and Biotechnology
Oak Ridge National Laboratory
davisonbh@ornl.gov

For Presentation at the AIChE 2008 Annual Meeting, Philadelphia, November 2008

Extended Abstract

The challenge of converting sustainable cellulosic biomass into fuels has the opportunity for science and technology making an appreciable national impact in the next 20 years. Bioenergy interest is driven partly by the economics and national security. The effective conversion of lignocellulose into fermentable sugars is the dominant obstacle to cost-effective production of biofuels in sustained quantities capable of impacting U.S. consumption of fossil transportation fuels. Three key questions are: can we do this effectively and efficiently; can we make a significant impact; and can we do this sustainably? The application of new tools from modern system biology has the potential to overcome many current technological and economic challenges. We will discuss the biological approaches being taken to address these questions including the BioEnergy Science Center (BESC) at the Oak Ridge National Laboratory.

The author will touch on the converging drivers of climate and security which have led interest in biofuels to greatly increase since 2005. Bioenergy is a complex multiple choice problem where choices can be made in land, feedstock, process and product (or fuel). There are also several "generations" of technology competing for research and deployment support. This ranges from the "established" corn starch ethanol plants to the first generation cellulosic biorefineries, which are just breaking ground, to second generation technologies entering piloting on to third and fourth generation technologies in the laboratory. We are seeing a compressed technology development cycle for a commodity industry due to the important drivers.

Challenges for research and development extend into the supply chain and how biology can help find solutions (see Ragauskas *et al.*, 2006; and DOE Report "Breaking the Biological Barriers to Cellulosic Ethanol"). Part of the increased funding in this area has come from the U.S. Department of Energy's Office of Science through its Genomics: GTL program (http://genomicsgtl.energy.gov/) establishing of three fundamental Bioenergy Research Centers in late 2007. One of these, BESC (http://www.bioenergycenter.org/), is a multi-partner project pursuing an integrated strategy to understand and overcome biomass recalcitrance and provide accessible sugars. We believe that access to the sugars in lignocellulosic biomass is the current critical barrier for widespread adoption of biofuels and that solving this will cut processing costs significantly and be used in most conversion processes (see Lynd *et al.*, 2008). This includes all fermentation based processes whether to ethanol, butanol or other potential fuels.

BESC is following a two-pronged approach to increase the accessibility of biomass sugars: modify the plant cell wall structure to increase accessibility and improve combined microbial approaches that release sugars and ferment into fuels. This can be visualized in Figure 1 where a current conceptual baseline for a generation one cellulosic biorefinery is shown along with a potential future concept. The BESC is using a strategy of developing various integrated higher-throughput pipelines for generating and screening large numbers of samples of modified biomass or biocatalysts followed by more detailed analysis of the most interesting ones.

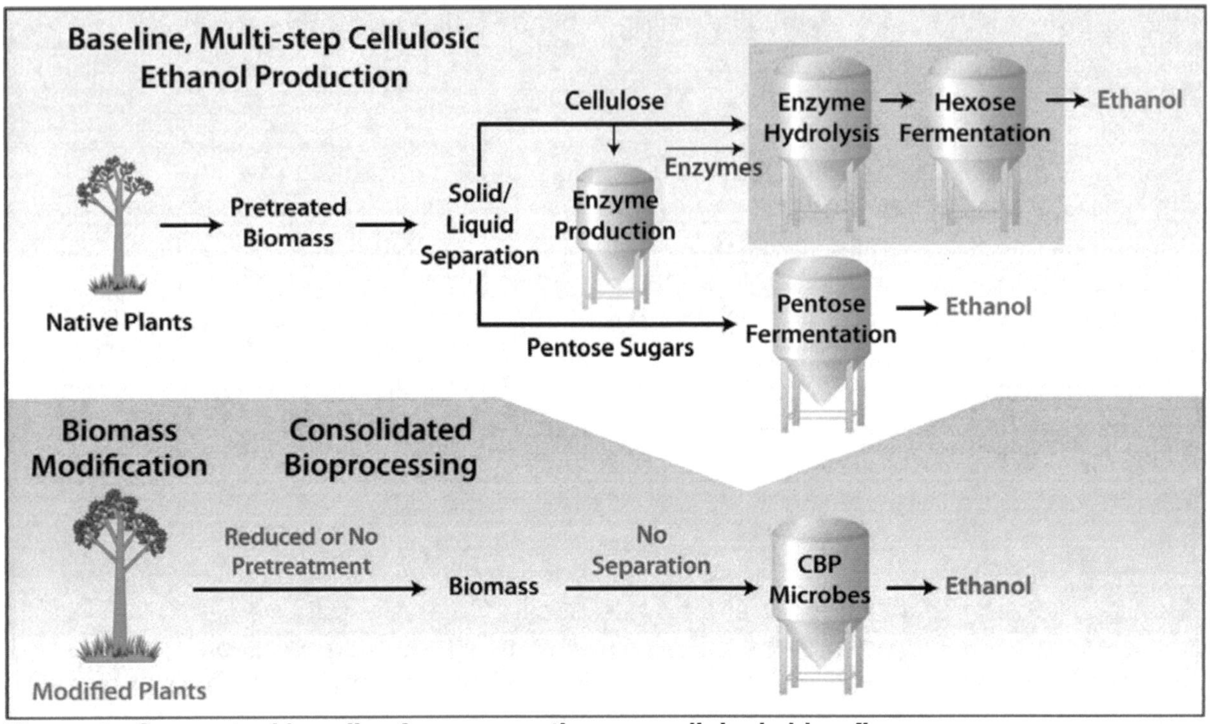

Figure 1: Conceptual baseline for a generation one cellulosic biorefinery.

References

Lynd, L.R., M.S. Laser, D. Bransby, B.E. Dale, B. Davison, R. Hamilton, M. Himmel, M. Keller, J.D. McMillan, J. Sheehan, C.E. Wyman, "How Biotech Can Transform Biofuels," *Nature Biotechnology* **26**:169-172 (2008).

Ragauskas, A.J., C.K. Williams, B.H. Davison, G. Britovsek, J. Cairney, C.A. Eckert, J. Frederick, J.P. Hallett, D. Leak, C.L. Liotta, J.R. Mielenz, R. Murphy, R. Templer, T. Tschaplinski, "The Path Forward for Biofuels and Biomaterials," *Science* **311**(5760):484-489 (2006).

DOE Report "Breaking the Biological Barriers to Cellulosic Ethanol," can be found at http://genomicsgtl.energy.gov/biofuels/.

Acknowledgement

The BioEnergy Science Center is a U.S. Department of Energy Bioenergy Research Center supported by the Office of Biological and Environmental Research in the DOE Office of Science.

450c Biofuels: Unlocking the Potential

Joseph Kocal and Jennifer Holmgren. UOP, 50 East Algonquin Road, Des Plaines, IL 60017

Government policy is driving substitution of petroleum feedstocks by biologically derived feedstocks. However, as biofuels become increasingly pervasive, the debate over the potential of biofuels to become a sustainable source of liquid transportation fuels continues. Much of this debate is accentuated by life cycle analyses of biofuel production routes, including low yields of biofuels from crops that would otherwise be used for food production, the poor energy balance for producing some biofuels and a lack of production and distribution networks. Ameliorating these risks are the intensive efforts underway to develop technologies that can process 2nd generation feedstocks such as agricultural residues and forest wastes as well as inedible and algal oils. The long term sustainability of a biofuels industry will be largely driven by the availability of these 2nd generation feedstocks and the technologies to process them.

450d Poet's Vision for the Future of Biorefining

Jason Kwiatkowski and David Carlson. POET Research, POET, 4615 N. Lewis Ave., Sioux Falls, SD 57104

The Energy Independence and Security Act of 2007 was signed into law on December 19, 2007. Included in this act was an amendment to the Renewable Fuels Standard (RFS) that makes provision for 36 billion gallons of renewable fuel by 2022, 16 billion gallons of which are specified to be cellulosic biofuel. As the largest ethanol producer in the United States and in the world POET is committed to meeting these ambitious targets.

This presentation will discuss POET's strategies towards meeting the goals of the RFS, including both improved starch to ethanol and biomass to ethanol. Additionally, POET's vision of an integrated biorefinery that moves beyond liquid fuel production will also be presented.

450e Process Engineering of Renewable Resources for Ethanol Production

Michael Ladisch, Purdue University, Department of Agricultural & Biological Engineering, 1146 ABE Building, West Lafayette, IN 47907-1146

The production of ethanol from cellulose for use as a liquid transportation fuel requires a combination of process engineering, microbiology, and accessibility to feedstock. The feedstock must be available to supply the plant 24 hours / day, 7 days per week. Siting of the plant is key to ensuring feedstock supply. Conversion of the feedstock to sugars and to ethanol requires pretreatment, hydrolysis, and fermentation. Pretreatment softens up the plant cell wall structure and enables enzymes to access the cellulose so that they may catalyze the formation of monosaccharides. The monsaccahrides, in turn, may be converted to ethanol through microbial fermentation by yeast or bacteria that have been engineered to convert both glucose and xylose to ethanol. During the bioconversion steps, cascading molecular control of enzyme activity occurs due to inhibitors that are formed during the pretreatment and/or hydrolysis steps. This paper discusses the role of process engineering in addressing issues of inhibition, solids loading, and fermentation, and gives a review of fundamental mechanisms and future research needs for converting renewable resources to biofuels in a cost effective manner.

460a Defining Enzyme Inhibitors for Cellulose Hydrolysis

Eduardo Ximenes, Laboratory of Renewable Resource Engineering (LORRE), Purdue University, Potter Engineering Center, 500 Central Dr., West Lafayette, IN 47906, Youngmi Kim, Laboratory of Renewable Resources Engineering (LORRE), Purdue University, Potter Engineering Center, 500 Central Drive, West Lafayette, IN 47907-2022, Nathan S. Mosier, Agricultural and Biological Engineering & Laboratory of Renewable Resources Engineering (LORRE), Purdue University, Potter Engineering Center, 500 Central Drive, West Lafayette, IN 47907-2022, and Michael R. Ladisch, Laboratory of Renewable Resources Engineering Department, Purdue University, Potter Engineering Center, 500 Central Drive, West Lafayette, IN 47907-2022.

Compounds that inhibit the enzymatic hydrolysis of cellulose to fermentable sugars may be present when lignocellulosic materials are pretreated. We have observed the reduction in extent of conversion of pretreated cellulose to fermentable sugars as the solids loading (concentration of solid biomass material) is increased even though the ratio of enzyme to solids material is kept constant. In addition, washing the pretreated cellulosic material results in greater extents of conversion. We have identified various potential inhibitors including ferulic acid and other phenolic compounds which may help to describe this phenomenon. Glucose yield reduction up to 70% for the hydrolysis of wet cake by commercial cellulase and ß-glucosidase in the presence of 0.5% (w/v) ferulic acid were observed in our studies. Comparing enzymes from different microorganism sources, we have also found that some are more resistant to inhibition by ferulic acid than others. This paper describes the chemistry and possible mechanisms for inhibition of industrial cellulase and hemicellulase enzymes by unknown components that are present in lignocellulosic material.

460b Production of Biofuels by Simultaneous Utilization of Mixed Sugars

Hideaki Yukawa, Molecular Microbiology and Biotechnology Group, RITE, 9-2 Kizugawadai, Kyoto, 619-0292, Japan

Biorefinery R&D has progressed dramatically as a result of recent advancements in the field of biotechnology, with industry in Japan realizing the concept as a novel technology which could lead to the establishment of a new industry.

As current corn-based bioethanol production cannot satisfy the growing demand for bioethanol, tomorrow's production technology, in which ethanol is produced from cellulosic biomass, must be developed. The new technology must achieve high yields, be capable of simultaneous utilization of pentose and hexose sugars, and be resistant to fermentation inhibitors derived from lignocellulose degradation.

The RITE bioprocess is based on a novel concept that enables highly efficient production of a variety of biochemicals using microbial cells whose growth is artificially arrested. In essence, the growth-arrested, genetically modified Corynebacterium glutamicum cells used in the bioprocess function like inorganic catalysts of chemical processes. In the RITE bioprocess therefore, it is possible to fill a reactor to a high density with microbial cell catalysts and continuously supply a biomass substrates (sugars) through the reactor to carry out the reaction at high speed. As a result, productivity (space time yield; STY) of the RITE bioprocess is much higher than that of conventional bioprocesses and is comparable to that of chemical processes.

In collaboration with Honda R&D Co. Ltd. we aim for early industrial application of the RITE bioprocess in bioethanol production. To this end, a pilot plant built in the premises of Honda R&D Co. Ltd. should provide the technical data requisite for industrial scaling up.

In this presentation, current advances on biofuel and biochemical production at RITE will be discussed.

460c Enzymatic Hydrolysis of Afex Treated Corn Stover by Cellulolytic and Hemicellulolytic Synergistic Enzyme Cocktails

Shishir Chundawat[1], Dahai Gao[1], K. Chandraraj[2], John Poland[3], Justin Stege[3], Mary Lipton[4], Venkatesh Balan[1], and **Bruce E. Dale**[1]. (1) Chemical Engineering and Materials Science, Michigan State University, 2527 Engineering Building, East Lansing, MI 48824, (2) Department of Biotechnology, Indian Institute of Technology (Chennai), Biosciences Building, Chennai, 600036, India, (3) Verenium Corporation, San Diego, CA 92121, (4) Biological Sciences, Pacific Northwest National Laboratory, Richland, WA 99352

The identification of a suitable cocktail of cellulases and hemicellulases to hydrolyze residual glucan and xylan becomes pertinent in the case of lower severity pretreatments (e.g. ammonia fiber expansion or AFEX). A fundamental understanding of enzyme synergy in the hydrolysis of pretreated biomass requires a high-throughput combinatorial enzyme screening approach for lignocellulosic biomass.

Standard enzymatic hydrolysis protocols for cellulosic enzyme systems have several inherent disadvantages including long analysis times, excessive reagent and substrate usage, high labor inputs and non-realistic substrates (e.g. filter paper, purified xylans, and chromogenic substrates). The choice of an optimum enzyme cocktail depends largely on the substrate characteristics rather than standard enzyme-activities that are currently measured. Screening multi-enzyme systems directly on pretreated lignocellulosics would be a better way of identifying optimum synergistic cocktails of enzymes. The automated 96-well BCRL microplate method is a rapid hydrolytic assay technique (Chundawat et al., 2008, Biotech Bioeng, 99, 6:1281-1294) that is currently employed at our lab.

Commercially available cellulase (Spezyme® CP) was purified to isolate cellobiohydrolase (CBH) I, CBH II and Endoglucanase (EG) I. Ternary mixtures of purified cellulases were used to hydrolyze AFEX treated corn stover to determine optimum ratio of CBH I: CBH II: EG I. It was found that CBH II does not have significant synergistic interaction with either CBH I or EG I. AFEX corn stover was also hydrolyzed by several mixtures of hemicellulases and other accessory enzymes from Genencor and Novozyme in the presence of well defined cellulase cocktails (e.g. Accellerase® and purified CBH I/EG I/beta-glucosidase) to maximize synergistic hydrolysis yields with minimum enzyme loading (0.5-30 mg/gm glucan). Individual protein components for several commercial enzymes were identified using a high-throughput proteomics (LC-MS/MS) approach. Purified Aspergillus nidulans beta-xylosidase and xylanase were also isolated to develop a rational, minimal enzyme cocktail for AFEX treated corn stover. Several proprietary cellulases and hemicellulases from Verenium Corporation were also screened on AFEX corn stover.

The overall goal of the current project is to develop an enzyme cocktail specifically tailored for AFEX treated lignocellulosics to help reduce total protein loading employed during enzymatic hydrolysis.

Web Page: www.everythingbiomass.org

460d Evaluation of Cellulose Reactivity with Purified Enzyme Components during Hydrolysis

Jian Shi, College of Engineering-Center for Environmental Research and Technology, University of California, Riverside, 1084 Columbia Ave, Riverside, CA 92507, Bin Yang, University of California Riverside, 1084 Columbia Avenue, Riverside, CA 92507, Edmund Larenas, Genencor, A Danisco Divison, 925 Page Mill Rd.,, Palo Alto, CA 94304, Colin Mitchinson, Genencor International, 925 Page Mill Road, Palo Alto, CA 94304, and Charles E. Wyman, College of Engineering-Center for Environmental Research and Technology, University of California Riverside, 1084 Columbia Avenue, Riverside, CA 92507.

To better understand the mechanism of enzymatic hydrolysis of cellulose, we applied a restart protocol to explore how cellulose reactivity changed and the interaction of cellulose with major cellulase components over the course of enzymatic hydrolysis. This approach allowed us to accurately monitor the hydrolysis rate of partially converted cellulose over the time course of substrate digestion excluding the impact of cellulase changes. In this presentation, the effect of enzyme-substrate interactions on reaction rates was studied using purified key cellulase components (i.e., CBHI, EGI, and/or EGII) from wild type Trichoderma reesei, and their synergism was studied by comparing the interrupted enzymatic hydrolysis of pure cellulose as measured using the restart protocol with uninterrupted hydrolysis as well as with the hydrolysis behavior of individual key cellulase components. Important chemical and physical features, including DP, reducing ends, crystallinity index, accessible surface area, and cellulose reactivity with enzymes, were characterized over the time course of cellulose hydrolysis, and the effects of cellulose reactivity and cellulase processivity on enzymatic hydrolysis of cellulose will be discussed.

460e Understanding and Harnessing the Microbial Fermentation of Glycerol. A New Path for the Production of Biochemicals

Ramon Gonzalez, Departments of Chemical & Biomolecular Engineering and Bioengineering, Rice University, MS-362, P.O. Box 1892, Houston, TX 77251-1892

Glycerol is an abundant, inexpensive, and highly reduced molecule generated as inevitable by-product of biofuels production. The use of glycerol as carbon source in fermentation processes offers the opportunity to produce reduced chemicals at yields higher than those obtained with the use of common sugars. Fully realizing this potential, however, would require the metabolism of glycerol in the absence of external electron acceptors. Unfortunately, fermentative utilization of glycerol is restricted to a small group of microorganisms, most of them not amenable to industrial applications. The synthesis of 1,3-propanediol (1,3-PDO) by these organisms has long been considered the metabolic property that determines their ability to ferment glycerol. For example, *E. coli* and *S. cerevisiae*, two microorganisms considered workhorses of modern biotechnology, are unable to synthesize 1,3-PDO and therefore thought to metabolize glycerol only via a respiratory process.

We have recently discovered that *E. coli* can indeed ferment glycerol in a 1,3-PDO-independent manner. We identified 1,2-propanediol (1,2-PDO) as a fermentation product and established the pathway that mediates its synthesis as well as its role in the metabolism of glycerol. Based on our findings, we proposed a new model for the fermentative utilization of glycerol in which: (1) the production of 1,2-PDO provides a means to consume reducing equivalents generated in the synthesis of cell mass, thus facilitating redox balance, and (2) the conversion of glycerol to ethanol, through a redox-balanced pathway, fulfills energy requirements by generating ATP via substrate-level phosphorylation. Other auxiliary or enabling pathways facilitating this metabolic process were identified along with the culture conditions triggering them, and therefore facilitating glycerol fermentation. We have also shown that this model for the 1,2-PDO-dependent fermentation of glycerol is valid for other microorganisms.

The knowledge base created by the aforementioned studies has been instrumental in the implementation of metabolic engineering strategies to convert glycerol to a variety of products, including lactate, succinate, 1,2-PDO, formate, and ethanol. We will present at the meeting our latest results in the understanding of this new metabolic competency and the metabolic engineering of these microorganisms to produce different chemicals from glycerol.

References:

Dharmadi, Y., Murarka, A., Gonzalez, R. (2006). Biotechnol. Bioeng. 94: 821-829.

Yazdani. S. S., Gonzalez, R. (2007). Curr. Opin. Biotechnol. 18: 213-219.

Murarka. A., Dharmadi, Y., Yazdani, S. S., Gonzalez, R. (2008). App. Environ. Microbiol. 74: 1124-1135.

Gonzalez, R. Murarka. A., Dharmadi, Y., Yazdani, S. S. (2008). Metab. Eng. (In Press).

Yazdani, S. S., Gonzalez, R. (2008). Metab. Eng. (MS in Review).

460f Improve Hydrolysis Rate by Displaying Multi-Complex of Cellulolytic Enzymes and Produce Ethanol In One-Step Process

Seunghyun Ryu, Department of Chemical Engineering, Texas Tech University, MS 43121, Lubbock, TX 79409-3121 and M.Nazmul Karim, Chemical Engineering Department, Texas Tech University, MS 43121, Lubbock, TX 79409-3121.

Simultaneous saccharification and fermentation (SSF) steps have been performed in order to reduce the cost of bioethanol production process. A whole-cell biocatalyst system was developed previously in our system to directly produce ethanol from cellulose in a single step. Three types of cellulases, which were endoglucanase, cellobiohydroalse and beta-glucosidase originated from Clostridium cellulolyticum were expressed on the cell surface. Genetically engineered Escherichia coli LY01 (Dr. LO Ingram, University of Florida) was used as a host microorganism. Our previous system was shown that the hydrolysis rate was 83 % and the theoretical ethanol yield was 87 % from amorphous cellulose. Attempts to improve the saccharification efficiency, we construct multi-complex of cellulases on the surface of microorganism.

Sustainable Biofuel Production: A DOE Perspective

__Alison M. Goss Eng__, Energy Efficiency and Renewable Energy, Office of the Biomass Program, United States Department of Energy, 1000 Independence Avenue, SW, Washington, DC 20585

Biomass research has been a cornerstone of the Department of Energy's (DOE's) renewable energy research, development and deployment efforts over the last 25 years. In order to encourage the economic livelihood of a thriving biofuel industry, the Department of Energy's Office of the Biomass Program (OBP) supports research and development aimed at assessing the impacts of biofuels on the environment, including impacts to land, water, and air from energy production and use. Included in this mission is a goal to substantially reduce greenhouse gas emissions by accelerating the adoption of renewable energy technologies.

A clear driver of the OBP's activities is the President's goal to increase the use of biofuels and other alternative fuels in the transportation sector in order to replace 20% of the gasoline demand in the United States by 2017 (referred to as the "20 in 10" goal). Meeting this goal will require: significant and rapid advancements in biomass feedstock and conversion technologies; availability of large volumes of sustainable biomass feedstock; demonstration and deployment of large-scale integrated biofuels production facilities; and biofuels infrastructure development efforts. In addition, the existing agricultural, forestry and commercial sectors will be making decisions to invest in biomass systems—from shifting land use, to building capital-intensive biorefineries, to establishing the infrastructure and public vehicle fleet for ethanol distribution and end use—in the context of economic viability (including as it relates to environmental sustainability) and the needs of the marketplace.

Figure 1. The Department of Energy's Biomass Program is researching the environmental impacts of biofuel production across the supply chain from feedstock production to end use in vehicles.

OBP's research and development (R&D) has led the effort to develop technology necessary to sustainably produce, harvest, and convert a variety of biomass feedstocks, as well as to deploy the resulting biofuels. Core R&D on feedstock production and logistics and biomass conversion technologies is conducted to develop the scientific and technical foundation that will enable the new bioindustry. OBP is looking to advance science in these areas through important collaborations with the DOE Office of Science Bioenergy Centers, the U.S. Department of Agriculture (USDA), land grant universities, and private industry. OBP has developed Regional Feedstock Partnerships to begin to realize the sustainability of the resource potential outlined in the "Billion Ton Study"[1]. This approach facilitates the collaboration of industry, the agricultural community, state and local governments and USDA and is expected to accelerate the resource-readiness as the cellulosic fuels industry emerges.

The core R&D of OBP is organized around the integrated biorefinery concept. The biorefinery helps deliver sustainable and environmentally sound contributions to power, fuels, and products demand while supporting rural economies. Key barriers relevant to this area include ensuring resource sustainability at levels large enough to support large-scale production facilities and maximizing the efficiency of conversion facilities to minimize costs. Energy production from biomass on a large scale will require careful evaluation of U.S. agricultural resources and logistics, as these will likely require a change in paradigm that will take time to implement. Current harvesting, storage and transportation systems are currently inadequate for processing and distribution of biomass on the scale needed to support dramatically larger volumes of biofuels production. Evaluating the current feedstock resources on a national level as well as the potential for future feedstock production in light of environmental constraints is part of OBP's focus.

The primary program areas related to ecosystems services and sustainability within DOE OBP are (1) sustainable feedstock production, (2) sustainable harvest, and (3) sustainable biofuels production. The program is also committed to engaging in the international dialogue about sustainable biofuel production and providing outputs from our research investments to support the discussion when appropriate.

Sustainable feedstock production

Existing data on the environmental effects of feedstock production and residue collection are not adequate to support lifecycle analysis of biorefinery systems. It is difficult to assure that residue biomass will be collected in a sustainable manner due to the lack of decision support tools to predict the effects of residue removal as a function of soil type. There is an additional lack of a selective harvest technology that can evenly remove only desired portions of the residue. Until the residue issue is addressed, particularly with regard to corn stover, the full utilization of agricultural residue for bioenergy will be severely constrained. The production and use of perennial energy crops also raise a number of sustainability questions (such as water and fertilizer inputs, establishment and harvesting impacts on soil, etc.) that have not been comprehensively addressed. The Biomass Program at the Department of Energy strives to address some of these issues through our sustainability research and development portfolio.

A central focus of feedstock production efforts is to establish and maintain Regional Biomass Energy Feedstock Partnerships in collaboration with USDA, the Sun Grant Initiative universities, and other regional partners. Collaborating in this manner will be crucial to overcoming specific geographic issues of varying climatic conditions, soil types, water quality, and land usage. These regional partnerships are necessary for assessing and quantifying the feedstock resource base because of the diversity of regional feedstocks in terms of growth requirements, climatic differences, and infrastructure needs. Work is focused on:

---

[1] Biomass as a Feedstock for Bioenergy and Bioproducts Industry: The Technical Feasibility of a Billion-Ton Annual Supply, Robert D. Perlack, et al., USDA/DOE, DOE/GO-102005-2135, April 2005.

1) Genetic improvement of crops including yield per acre as well as traits desirable for conversion to biofuels, an effort conducted primarily by USDA, DOE's Office of Science, land grant universities, and private companies;
2) Regional resource assessments of the types of biomass feedstocks that can be sustainably grown in specific locations across the U.S., including the development of a National Bioenergy GIS;
3) Development of replicated field trials across regions to determine the impact of agricultural residue removal and to evaluate the feasibility of various energy crops; and
4) Analysis of the sustainability of producing specific biomass feedstocks, an effort being addressed by the Regional Feedstock Partnerships, USDA, and OBP analysis efforts.

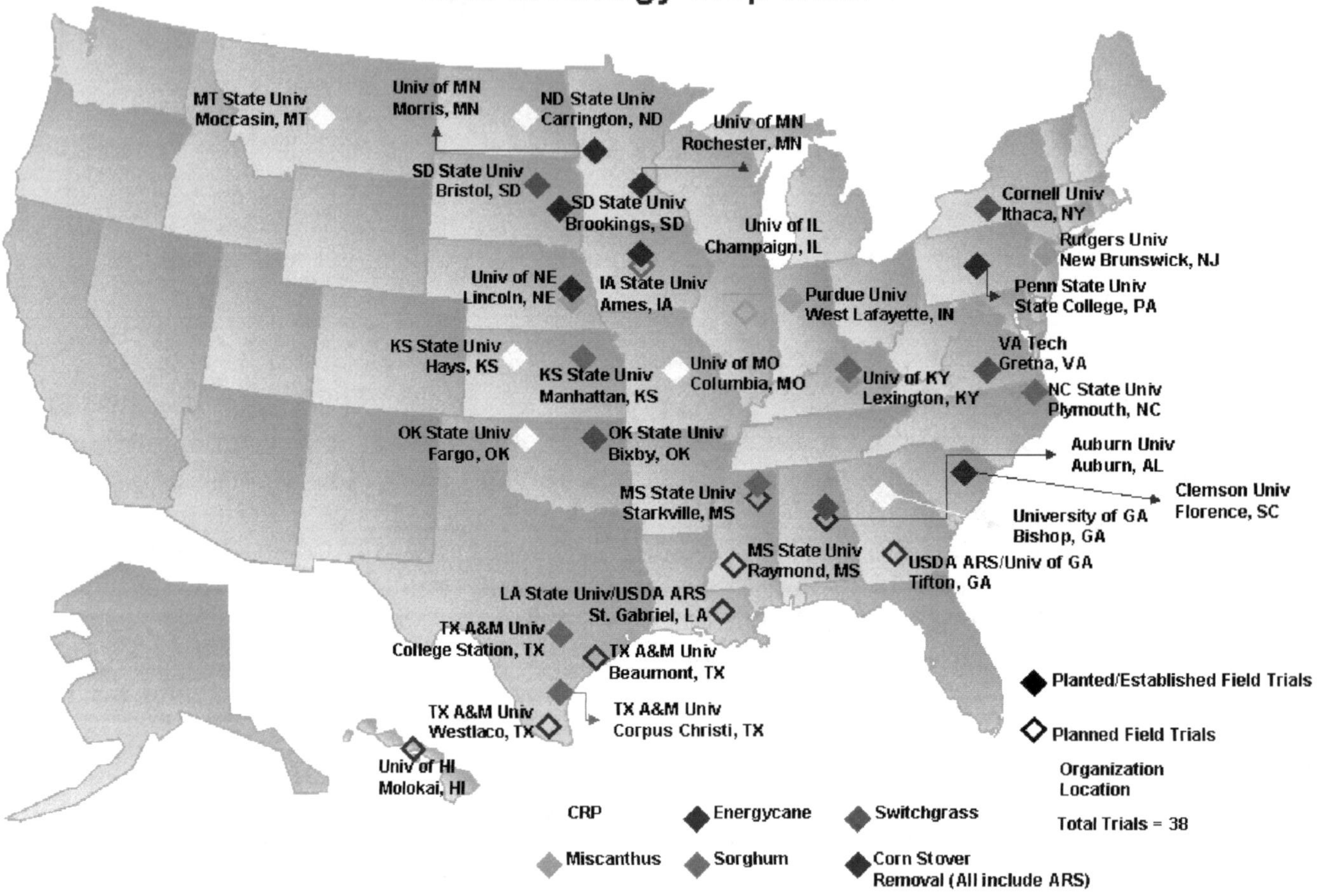

Figure 2. Several of the Regional Biomass Energy Feedstock Partnership Field Trials are focused on evaluating the impact of growing energy crops on soil carbon, water use and nutrient cycling.

One focus of the Regional Feedstock Partnership trials is evaluating the impact of switchgrass production on hydrology and water quality and on soil carbon sequestration. The hydrological and water quality impacts of land conversion to grow switchgrass have not been sufficiently studied. In particular, there is a need for studies that use the small watershed as the experimental unit. Test sites have been established in South Dakota, Oklahoma, Alabama, and New York to capture a wide potential distribution of switchgrass production. In regards to soil carbon sequestration, the relationship between it and residue removal is a potential area of concern with wide-scale production of bioenergy feedstocks. The objective for focusing on this area would be to apply the soil carbon balance model "CQESTR" to analysis of advanced agriculture residue removal scenarios for ethanol production, and implement the model as part of an enterprise level case study analysis of a proposed residue removal tool framework.

Additionally, the National Bioenergy GIS tool being developed by Oak Ridge, Argonne and Idaho National Laboratories and partnering universities, will serve as a spatially referenced database of current and potential feedstock availability and associated environmental and industrial variables to be used in the analysis of future economic and environmental sustainability related to feedstocks. This tool will also serve as a decision support system to inform the location of new feedstock production and processing facilities and to evaluate the resulting contribution potential of biofuels to the "20 in 10" goal (and beyond) that will sustain air and water resources of quality and availability for desired uses.

At a regional scale, we are studying the economic, energetic, logistic and environmental benefits of integrating the growth of short rotation woody crops into the current agricultural production areas. This effort would result it marginal land and water that are utilized to maximize the supply of additional woody lignocellulosic feedstock in parallel to corn ethanol or other crops, with concurrent optimization of water resources and environmental benefits. We have configured our study format to match existing ongoing work in OBP to examine feedstock growth potentials and water requirement implications for biofuel production. Within this framework, we are addressing a specific set of problems by addressing the potential use of impaired water and land that could be leveraged to meet OBP feedstock production goals.

We are integrating our research with an understanding of larger water utilization issues instead of a specific focus on impaired water and land use, along with examining the potential advantages of a more practical approach on yield potentials and on the minimization of the environmental footprint of current and future feedstock cropping.

Through a pilot assessment at the regional scale focused on the availability of marginal land and non-point nitrate pollution in groundwater and surface water, future work will include: determining quantitative estimates of potential growth of biofuels in "marginal land" by category; efficiency and usefulness of using impaired, nitrate-contaminated water to grow bioenergy crops, the logistic feasibility of this approach, with an analysis of the potential economic, environmental and energy gains; and impacts on surface and groundwater resource inventories, both clean and impaired.

Sustainable harvest

Current crop harvesting machinery is unable to selectively harvest desired components of biomass and address the soil carbon and erosion sustainability constraints. Biomass variability places high demand and functional requirements on biomass harvesting equipment. Current systems cannot meet the capacity, efficiency, or delivered price requirements of large cellulosic biorefineries, nor can they effectively deal with the large biomass yields per acre of potential new biomass feedstock crops. In addition, feedstock specifications and standards against which to engineer harvest equipment, technologies, and methods, do not currently exist.

A key to success is the ability to convert a wider variety of regionally-available biomass feedstocks and agricultural waste. The DOE is working to establish Regional Biomass Energy Feedstock Partnerships that will identify local opportunities for feedstock development and ethanol production. Part of this work is the development of a number of different core models that examine the impacts of farming, residue/crop harvesting and supply system practices on previously identified agronomics indicators such as soil organic carbon, erosion and water usage. These models develop basic relationships between crop types, soil types, land allocation and cropping schemes to determine the impact on these indicators. This task focuses on capturing the relationships in the core models for a systems level analysis of the biomass sustainability system based on potential growth scenarios of biomass demand. It is aimed at evaluating the current models and databases that will be combined into resources assessment models with the capability to analyze the feedstock supply systems to ensure that the sustainable resources are not stranded from the biorefineries and that the sustainable biomass production is sufficient to meet Federal goals. The next step will be to couple the Idaho National Laboratory-developed Feedstock Supply System model with the aforementioned National Bioenergy GIS to allow for a comprehensive assessment of the cost of a biomass feedstock that includes the feedstock supply system costs.

The output from this research will be a system of models that will make it possible to do a comprehensive analysis of the entire Feedstock Supply Logistics system. By linking the INL Feedstock model with the National Bioenergy GIS, it will be possible to assess the impacts of the supply systems on the cost basis for the various feedstocks, an assessment that is missing in the current analyses. The outcome should include economic cost analysis and sustainability indexes for the various feedstocks under consideration.

Future work includes developing the capability to analyze the cost impacts of the supply systems based on variability in feedstock access. This analysis will help to identify potential sources of constraints on growth and help mitigate impacts of those constraints. The final analysis will determine the sustainability of the biomass supply system and what the constraining factors will be for growth.

Sustainable biofuels production

While perhaps the greatest sustainability challenges to biofuel production lie within the feedstock production sector of the biomass-to-bioenergy supply chain, existing projects within OBP extend beyond feedstock production and harvest. Currently, a life cycle assessment (LCA) of the Advanced Energy Initiative is being performed for the 60 billion gallon 30x30 scenario (a scenario for supplying 30% of 2004 motor gasoline demands by 2030). The analysis covers the entire biofuels supply chain from feedstocks to vehicles. The four main areas addressed in the LCA are: land use and soil sustainability, water use impacts, air quality impacts, and greenhouse gas (GHG) emissions impacts.

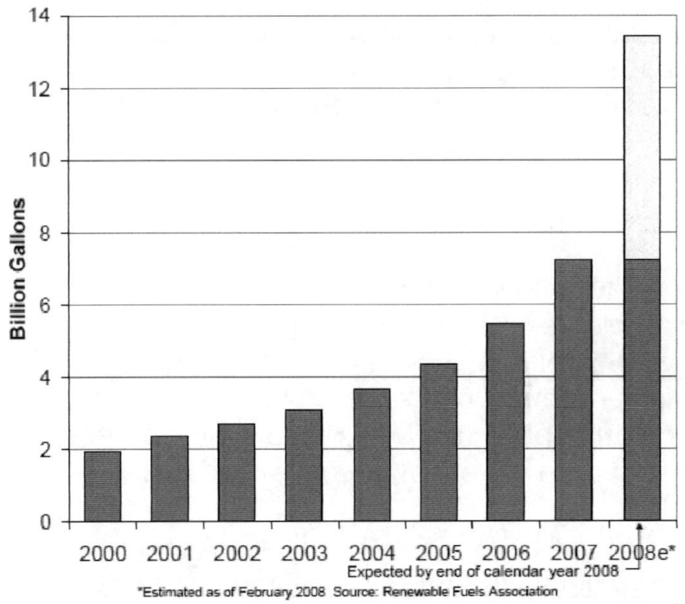

Figure 2. As U.S. ethanol production capacity increases, potential impacts to water use, air quality and greenhouse gases must be assessed.

Also, the GREET model (Greenhouse gases, Regulated Emissions, and Energy use in Transportation) is being utilized for an analysis of water demand for biofuel production, energy and GHG emission benefit of biofuels. Included in this project is an expansion of the existing model to include corn ethanol, sugarcane ethanol, and flex-fuel vehicle (FFV) test results. As part of GREET model development, a focus on indirect land use change is underway. Land-use practices that support biofuel feedstocks and are sustainable cannot be adequately defined due to the lack of documentation of indirect effects of bioenergy and the integration of land use into LCA models. The Biomass Program is engaging with its national lab and university partners to enhance existing analytical tools and develop new predictive models of land use implications of biofuel production. This will include preparing and updating databases on land use change, productivity, inputs, and carbon flows. These tasks will also include testing models with empirical data for validation and the identification of weaknesses and/or areas of improvement related to estimates of carbon loss and storage under different feedstock production systems.

International work

Through a partnership with the nonprofit environmental organization, Conservation International, the DOE is completing an in-depth analysis of potential areas for biofuel crop expansion and overlaying this information with regions with high value for biodiversity conservation, carbon storage, protection of water resources, or other ecosystem services. This work enables the identification of areas that should be avoided during biofuels expansion, as well as opportunities for growth onto abandoned or degraded lands. The development of industry-wide sustainability standards, as well as comprehensive national policies and regulatory frameworks for biofuel crop expansion will help ensure that producers follow recognized best practices by limiting market access to those that do not comply. This project is designed to help give policy makers, industry experts, and producers the tools they need to generate the socioeconomic benefits of biofuels while mitigating the potential negative impacts on biodiversity.

Projects such as that with Conservation International are further enabling DOE to provide data and analysis to inform a variety of international and domestic discussions on sustainability standards including the Roundtable on Sustainable Biofuels and the Global Bioenergy Partnership's working groups on Greenhouse Gas Accounting and Sustainability. In partnership with the State Department, DOE is working to address global sustainability issues with international partners, including environmental organizations, industry, and others.

Through the International Biofuels Forum (IBF) Brazil, China, India, South Africa, and the European Commission have joined the US in an international discussion to promote the sustained use and production of biofuels around the globe. We believe that global standards for sustainable development would promote adherence to best practices in developing biofuels industries.

Conclusions

Through its focus on the entire biomass to biofuel lifecycle, the Biomass Program at the US Department of Energy is committed to developing biofuels that minimize or eliminate the negative environmental impacts associated with conventional fuels and have less carbon emissions throughout their lifecycle—from biomass production to end use in vehicles. To produce these biofuels, we are developing the resources, technologies, and systems needed to *sustainably* produce, harvest, transport, and convert a variety of biomass feedstocks.

For more information, visit: http://www1.eere.energy.gov/biomass/

511b Sustainable Strategies for Biofuel Production

Tom L. Richard[1], Gustavo Camargo[1], Rob Anex[2], and Ed Van Ouwerkerk[2]. (1) Agricultural and Biological Engineering, The Pennsylvania State University, 225 Agricultural Engineering Building, University Park, PA 16802-1909, (2) Agricultural and Biosystems Engineering, Iowa State University, 3202 NSRIC Building, Ames, IA 50011-3020

Four strategies are analyzed for their ability to produce large quantities of biofuel feedstocks without negative impacts on food production or the environment. These include:

1. Organic wastes, including Municipal Solid Waste (MSW), food processing wastes, wood processing residues, livestock mortalities and manure.

2. Perennial crops (short rotation trees or grasses) on abandoned or marginal land, and as streamside buffers and roadside plantings.

3. Integration of energy crops with food crops to increase the productivity of existing agricultural land, without reducing food production, and with enhanced environmental outcomes. Strategies include a) cover crops (winter canola, winter barley, winter rye); and b) crop residues (corn stover, wheat straw, etc.) when harvested at rates that maintain or enhance soil carbon and limit erosion – with enhanced harvests possible in combination with the cover crops in example a).

4. Enhanced management of forest landscapes. Many U.S. forests are overstocked with small diameter non-timber species left over from generations of "high-grading" (taking out just the best timber species). Thinning is widely recognized as improving wildlife habitat and forest recreation as well as timber production. As with cover crops, forest thinning for biofuels can provide an economic incentive for environmental enhancement.

This paper examines these alternatives for productivity, environmental impacts, and energy and nutrient use efficiency, with particular emphasis on option 3, which has largely been overlooked in the public debate.

511c Biological Pretreatment of Lignocellulosic Biomass during Wet Storage and Its Impact on Downstream Biorefinery Processes

Megan Marshall[1], Tom L. Richard[1], Qin Chen[1], Irene Darku[1], Lynn Petzke[2], Corey Radtke[2], Allison Ray[2], Farzaneh Rezaei[1], and Deepti Tanjore[1]. (1) Agricultural and Biological Engineering, The Pennsylvania State University, 249 Agricultural Engineering Building, University Park, PA 16802, (2) Biotechnology, Idaho National Laboratory, P.O. Box 1625, Idaho Falls, ID 83415

Storage and pretreatment of lignocellulosic biomass are both significant challenges facing the growing biorefinery industry. In humid environments, ensilage is a traditional biomass storage technique, which could be used to preserve the feedstock prior to bioconversion into ethanol. Under moist, anaerobic conditions, microorganisms convert the soluble sugars of the ensiled biomass into organic acids, resulting in low pH and minimizing further biomass degradation. Ensilage, either alone or with added biocatalysts, may have benefits as a combined storage and biological pretreatment strategy which is less energy and cost intensive. The goal of this work is to quantify these possible benefits. Variables considered include: feedstock (switchgrass and corn stover), silage duration (7-21 d for stover, 21 d-1 yr for switchgrass), experimental scale (lab mini-silos and field bales), and treatment during ensilage (enzymatic and microbial). Enzymatic treatments consisted of cellulase, hemicellulase, and/or ligninase addition while microbial treatments included inoculation with white-rot fungi before or after ensilage. These combined storage and pretreatment strategies were evaluated by using silage samples in enzymatic hydrolysis or simultaneous saccharification and fermentation reactions to quantify benefits in terms of sugar and/or ethanol yields.

Biomass Feedstock Removal and Transport Optimization

Remegio B. Confesor Jr. , Oregon State University
Gerald Whittaker, Agricultural Research Service, USDA
Rolf Färe, Oregon State University
Shawna Grosskopf, Oregon State University

Introduction

The research in conversion of agricultural biomass into ethanol has gained momentum due to high price of petroleum, the demand to reduce greenhouse gas emissions, and the recent US government policy on reducing foreign oil dependence. However, research in production and supply technologies should also be done in conjunction with ethanol conversion development. One challenge is in the collection, packaging, storage, and transport processes of the materials from the field to biomass refineries. This paper integrates the above processes into an optimized, multi-objective economic production model. We present a method that employs Shephard's activity analysis output model that is similar to the Data Envelopment Analysis (DEA). Activity analysis has been used to measure the efficiency and profitability of many agricultural and industrial production processes. The method that we propose to apply models dynamic production processes and intermediate production, among many other components of a production process. Shadow prices for undesirable outputs (pollutants) can also be calculated. Activity analysis is implemented in a linear program to calculate the optimum combination of inputs (costs) and outputs (revenues) of production processes, and can be dynamically linked to biophysical models to include environmental effects of production. We describe the components of our integrated modeling system below.

Optimization of Multiple Objecitives

A multi-objective optimization problem (MOOP) is generally understood to contain a number of objective functions that are to be minimized. Following Deb (Deb, 2001), the general form of a MOOP is:

$$\begin{aligned} \text{Minimize/Maximize} \quad & f_m(x), & m = 1,2,\ldots,M\ , \\ \text{subject to} \quad & g_j(x) \geq 0, & j = 1,2,\ldots,J\ , \\ & h_k(x) = 0, & k = 1,2,\ldots,K\ , \\ & x_i^L \leq x_i \leq x_j^U & i = 1,2,\ldots,n\ . \end{aligned} \tag{1}$$

The MOOP consists of *M* objective functions, with *J* inequality constraints and *K* equality constraints. A solution *x* is a vector of *n* decision variables that are constrained by lower x_i^L and upper x_i^U boundaries.

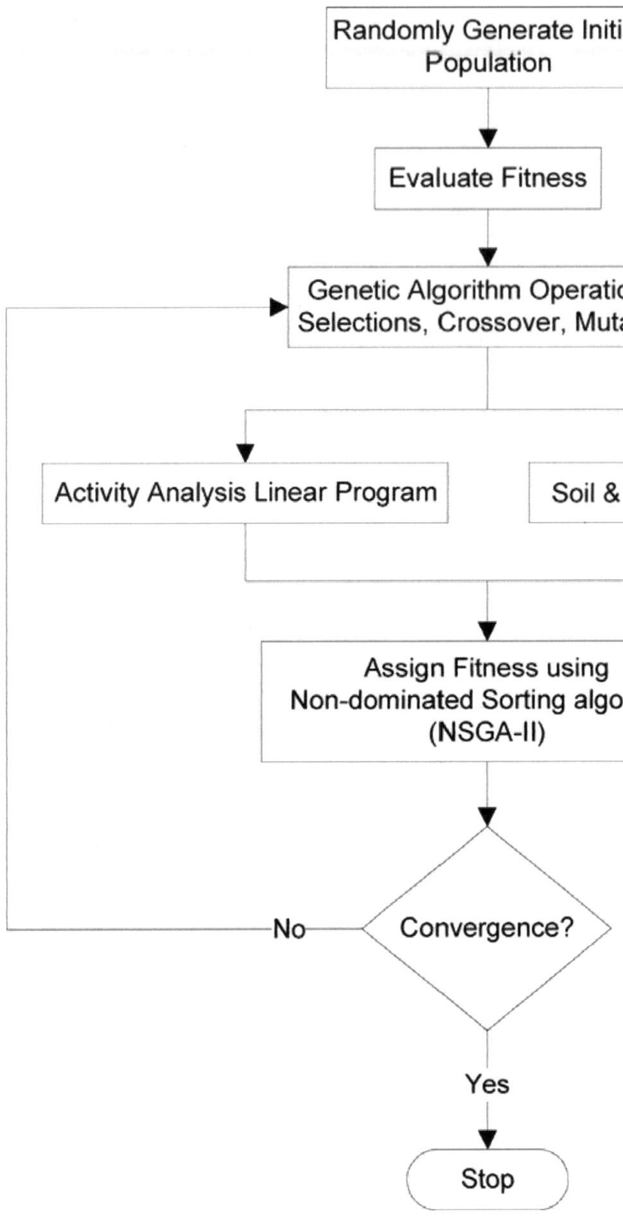

Figure 1. A hybrid genetic algorithm linking an economic model and a physical model (SWAT).

A simple genetic algorithm (GA) is an iterative algorithm based on retention of the best or "fittest" members of a population of answers until a stopping condition is satisfied (Goldberg, 1989). In an optimization application, the GA consists of an initial randomly generated population that is evaluated for fitness using an objective function, a test for convergence, and application of the GA operations of selection, crossover and mutation. These elements are followed iteratively until an optimum has been obtained (Figure 1). The non-dominated sorting algorithm (NSGA-II, Deb, 2002) is used to optimize over several objectives. This algorithm retains the non-dominated individuals from each generation, resulting in a estimate of the trade-offs among multiple objectives. The algorithm shown if Figure 1 uses and economic model to simulate producer behavior and a physical model to simulate the effect of producer behavior on the environment (Whittaker, el al., 2009). A novel feature of this algorithm is the dynamic linkage among models. Information is exchanged between models at each generation during the search for an optimal set. Almost all current integrated modeling systems optimize linked models in sequence, i.e., information only flows in one direction between the component models.

Economic Model – Activity Analysis

Each objective in the multi-objective genetic algorithm is specified by a model, or combination of models. In this study, profit maximization is assumed to be the objective of producers, and is optimized using activity analysis. In an activity analysis representation of a technology observed for a group of firms, there are $k = 1,\ldots,K$ decision making units (DMUs). Each DMU uses $x = (x_1,\ldots,x_M) \in \Re^M_+$ inputs to produce $u = (u_1,\ldots,u_N) \in \Re^N_+$ outputs. The observed inputs $x^k = (x^k_1,\ldots,x^k_M)$ and the observed outputs $u^k = (u^k_1,\ldots,u^k_N)$ are used together with the intensity variables $z^k \geq 0,\ k = 1,\ldots,K$, to form the reference technologies. Here we do not constrain the DMU profit, i.e., profit may positive or negative. We relax the assumption of constant returns to scale, and allow variable returns to scale by constraining the intensity variables to sum up to one. Our basic model is then

$$T = \left\{ (x,u):\ u_n \leq \sum_{k=1}^{K} z^k u^k_n, \qquad n = 1,\ldots,N, \right.$$
$$x_m \geq \sum_{k=1}^{K} z^k x^k_m, \qquad m = 1,\ldots M, \tag{2}$$
$$\left. \sum_{k=1}^{K} z_k = 1, \qquad z_k \geq 0 \right\}.$$

Denote input prices by $p^k \in \Re^M_+$ and output prices by $r^k \in \Re^N_+$. Then the profit of DMU k can be computed as the solution to the following linear program:

$$\pi(r^k, p^k) = \max \sum_{n=1}^{N} r^k_n u_n - \sum_{m=1}^{M} p^k_m x_m$$
$$s.t.\ \sum_{k=1}^{K} z^k u^k_n \geq u_n, \qquad n = 1,\ldots N,$$
$$\sum_{k=1}^{K} z^k x^k_m \leq x_m, \qquad m = 1,\ldots M, \tag{3}$$
$$\sum_{k=1}^{K} z_k = 1,$$
$$z_k \geq 0, \qquad k = 1,\ldots,K.$$

Additional constraints can be added to model optimization of transport from farm to a production facility.

Figure 2 illustrates this model with a technology consisting of 3 observations (a,b and c) of production where one input is used to produce a single output. The technology *T* is bounded by the line including a, b, and c, and includes the area under the boundary. The maximum profit is obtained by production at the facet where the price line is tangent to the boundary of the technology. There are three different input prices shown in Figure 2, p_1, p_2, p_3, while the output price remains constant.

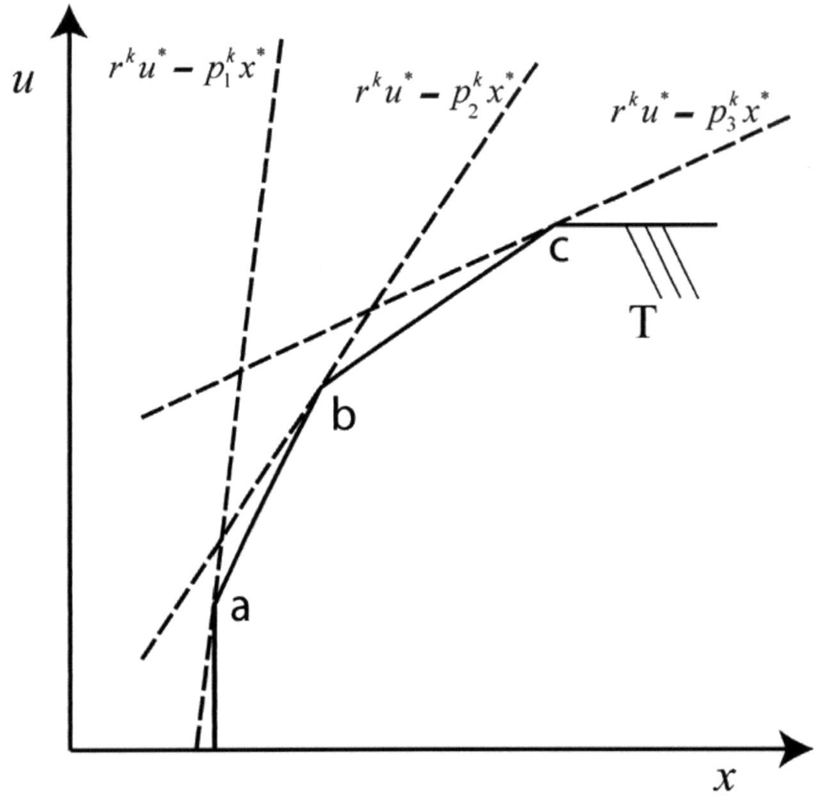

Figure 2. Profit maximization using Activity Analysis for technology T, facing three different input prices.

Physical Model

We use the Soil and Water Assessment Tool (SWAT) (Arnold et al., 1998) to simulate stream flow and associated movement of nitrogen, phosphorus, and sediment (among other variables) in the agricultural production of biofuel feedstock. SWAT is a geographically distributed model that simulates stream flow, plant growth and fate and transport of chemicals and sediment. In the genetic algorithm used for optimization, the profit maximizing amounts of inputs to production are calculated using activity analysis, and fed into SWAT to simulate the environmental effects of profit maximizing production.

Water Quality Index

In this study, we calculate the effect of agricultural production on water quality using the outputs from SWAT. An index gives a single number that represents the water quality of at the outlet of the simulated basin, and is calculated as the weighted sum of the metrics of water quality. These metrics include total suspended solids, levels of nitrogen and levels of phophorus, among others. By applying a variation of the Malmquist (1953) index, the weights are endogenous to the calculation. That is, they are derived from the observed data rather expert opinion , or some other method. We assume there are N characteristics of water quality x, and $k = 1,\dots,K$ measurements (observations) of water characteristics at $t = 1,\dots,T$ time periods. Our data is represented

$$x_k^t, \quad k = 1,\dots,K, \quad t = 1,\dots,T, \quad x_k^t \in \Re_+^N \quad . \tag{4}$$

In this study we define the reference technology (or benchmark) using all $K \times T$ observations, i.e., it is defined as

$$\begin{aligned} \Big\{ x = (x_1,\dots,x_N) : &\sum_{k=1}^{K}\sum_{t=1}^{T} z_k^t x_{kn}^t = x_n \qquad n = 1,\dots,N, \\ &\sum_{k=1}^{K}\sum_{t=1}^{T} z_k^t = 1 \\ &z_k^t \geq 0,\ t = 1,\dots,T,\ k = 1,\dots,K \Big\} \end{aligned} \tag{5}$$

where $z_k^t \geq 0$, are intensity variables. The intensity variables from the convex combinations of all the data points. The lower boundary of this set is the best practice benchmark isoquant II in Figure 1. For each observation (k',t) we compute

$$\begin{aligned} &[\theta_{k't}]^{-1} = \min \theta \\ &s.t. \quad \sum_{k=1}^{K}\sum_{t=1}^{T} z_k^t x_{kn}^t \leq \theta x_{k'n}^t \qquad n = 1,\dots,N \ , \\ &\qquad \sum_{k=1}^{K}\sum_{t=1}^{T} z_k^t = 1 \\ &\qquad z_k^t \geq 0,\ t = 1,\dots,T,\ k = 1,\dots,K \ . \end{aligned} \tag{6}$$

Suppose we compare (k',t) with $(k',t+1)$, then the water quality index (WQI) is

$$WQI_{k'} = \frac{\theta_{k',t+1}}{\theta_{k',t}} \quad . \tag{7}$$

In this application, we compare all measurements through all time periods, so $\theta_{k',t}$ is set equal to 1. If WQI = 1, water quality is equal to the best practice. WQI values greater than 1 indicate a poorer water quality. To implement the calculation, we use a linear program to solve (6) for each observation, then calculate (7), the water quality index.

For this study, we assume that the water characteristics are undesirable, and should be minimized. The two dimensional case is illustrated in Figure 1, where we wish to compare vector x^0 of length 0C with vector x^1 of length 0B. The benchmark that we use for comparison is the best practice benchmark II, where we are indifferent as to choice among points that lie on the benchmark. Each distance AB and DC, known as Shephard's input distance function, is the distance that each vector can be proportionally reduced to reach the benchmark (II). The ratio of the input distance functions, i.e. $\dfrac{OB/OA}{OC/OD}$ is the Malmquist consumer/input quantity index, and we conclude for this example that x^1 has better water quality than x^0.

Integrated Modeling System

Figure 4 shows the data flow of the complete integrated modeling system. The system is implemented in the statistical programming language R and FORTRAN. Given the large computational requirements of the modeling system, we set it up to run on a Beowulf cluster. The algorithm is parallelized by running the evaluation step of the GA for an individual on a node in the cluster.

The results from the use of this integrated modeling system may be used to assess the efficiency and economic viability of alternative technologies employed in agricultural biomass removal and transport for ethanol production. The system requires substantial knowledge to set up and run. We propose that the output of the modeling system distributed to stakeholders will be a database containing the set of trade-offs among objectives. Any stakeholder can find the optimum solution that fits their preferences by querying the database. Stakeholder choices will be informed by a state of the art modeling system, without a requirement to understand or run the modeling system.

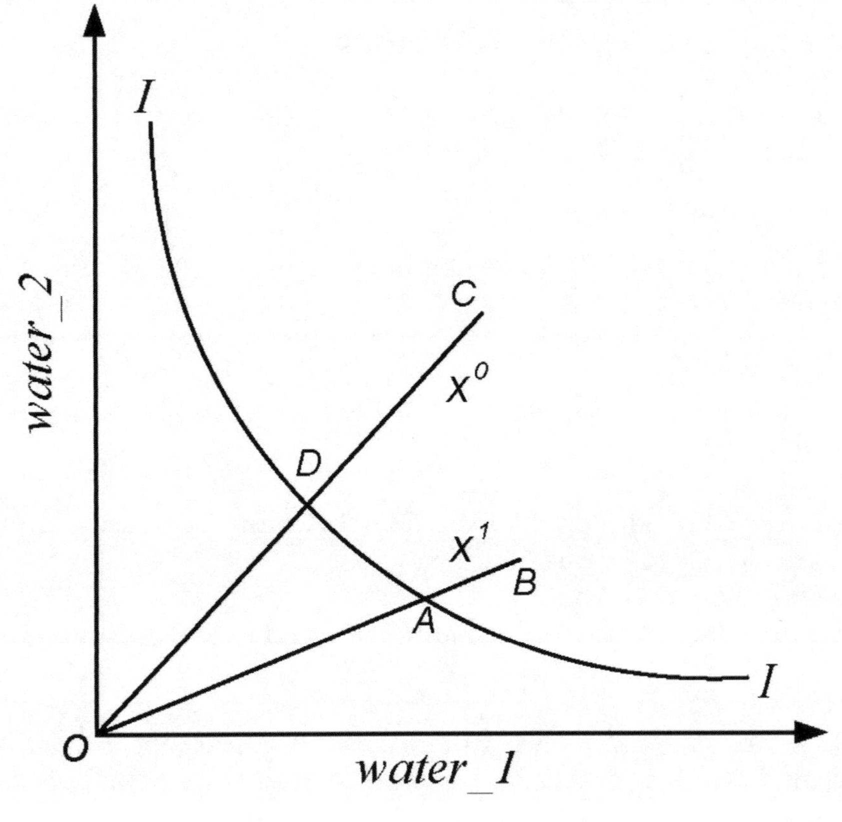

Figure 3. Construction of the Malmquist index with two constituents of water quality.

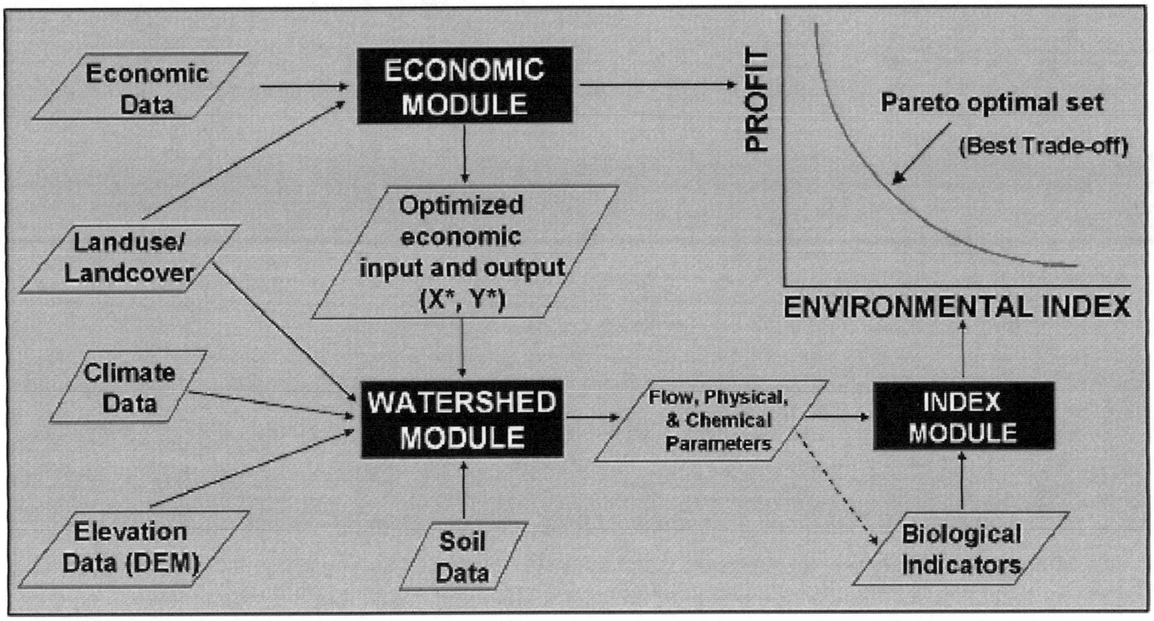

Figure 4. Information flow for integrated modeling system for multi-objective simulation of agricultural bio-fuel feed stock production.

References

Arnold, J.G., R. Srinivasan, R.S Muttiah, and J.R. Williams. 1998. Large Area Hydrologic Modeling And Assessment Part I: Model development. J. Am. Wat. Res. Asso. 34: 73-89.

Deb, K., 2001. Multi-Objective Optimization Using Evolutionary Algorithms. John Wiley & Sons, New York.

Deb, K., A. Pratap, S. Agarwal,and T. Meyarivan, 2002. A Fast And Elitist Multiobjective Genetic Algorithm: NSGA-II. IEEE Transactions on Evolutionary Computation, 6(2): 182.

Goldberg, D.E., 1989. Genetic Algorithms in Search, Optimization and Machine Learning. Addison-Wesley Longman, Inc., New York.

Malmquist, S., 1953, Index numbers and indifference surfaces, Trabajos de Estadistica, 4(1), 209.

Whittaker, G., R. Confesor, Jr., S.M. Griffith, R. Färe, S. Grosskopf, J.J. Steiner, G.M. Mueller-Warrant, G.M. Banowetz, 2009. A Hybrid Genetic Algorithm For Multiobjective Problems With Activity Analysis-Based Local Search. European Journal of Operational Research 193: 195–203.

511e Providing a Sustained Biomass Supply: Implications of Feedstock Variation

Timothy Rials, Director of R&D, Office of Bioenergy Programs, University of Tennessee, 2506 Jacob Drive, Knoxville, TN 37996-4570

Significant investment is being directed to the development of new technologies for converting lignocellulosic biomass to liquid fuels, primarily ethanol. Both thermochemical and biochemical processes are under development, and research is investigating the fundamental and technical challenges that currently impact economic viability. The equally important issue of feedstock production and supply for these biorefinery operations is receiving far less attention. With the planned construction of a pilot-scale facility, The University of Tennessee is concerned with providing feedstock for the 24/7 operation of this biochemical conversion plant. Recently, approximately 720 acres in east Tennessee (16 farms) have been seeded with switchgrass, with additional acreage planned for 2009. This operation provides a unique opportunity for study of the supply chain logistics associated with the sustained supply of a dedicated bioenergy crop. Also of interest is the question of feedstock quality. Near infrared spectroscopy has been developed for the high-throughput analysis of wood and biomass. Current work seeks to adopt this tool for on-line monitoring of important feedstock characteristics, including chemical composition. The potential to reduce variability of feedstock, or to more effectively utilize biomass property variation, will be discussed in terms of system viability.

511f Development of Feedstock Quality Specifications for Optimizing Biofuels Conversion Processes

Judy K. Partin, Materials Characterization, Idaho National Laboratory, 1765 N Yellowstone Highway, PO Box 1625, Idaho Falls, ID 83415-2211, Garold L. Gresham, Interfacial Chemistry, Idaho National Laboratory, 1765 N Yellowstone Highway, PO Box 1625, Idaho Falls, ID 83415-2208, and Peter A. Pryfogle, Biofuels and Renewable Energy, Idaho National Laboratory, 1765 N Yellowstone Highway, PO Box 1625, Idaho Falls, ID 83415-2203.

The DOE/USDA Billion Ton Technical Feasibility Study identified 1.3 tons per year of biomass feedstock that could potentially be used for biofuel conversion in the United States. One difficulty in the commercialization of this resource is that, unlike other major commodities, it is comprised of numerous smaller resources that have a variety of physical and chemical attributes. These attributes include material density and recalcitrance, carbohydrate, BTU, and lignin content, water activity, and the presence of various organic and inorganic compounds, or contaminants. Various collection, preprocessing, pretreatment, and storage operations may alter these properties in ways that impact the value of the feedstock for biochemical and thermochemical conversion processes. For example, harvesting could be managed to reduce moisture, influence carbohydrate and lignin content, or gather particular material fractions of the biomass. Wet storage and ensiling processes may present avenues for collecting soluble compounds, including sugars, salts, and other inorganic contaminates, or for reducing the recalcitrance of the biomass through pretreatment. Preprocessing operations, such as grinding and screening, could be used to reduce and select particle size distributions, resulting in higher material densities. In addition to increasing the biomass quality, these operations also present opportunities for reducing the handling, transportation, and storage costs of the feedstocks.

This paper describes on-going research directed at identifying the key feedstock attributes that impact conversion efficiencies, understanding how these properties can be changed during the feedstock assembly operations, and applying screening techniques to establish and track material quality through these operations. The goal of this work is to determine the quality factors for biomass feedstocks that could be used to select, value, and screen materials for use in specific conversion processes. In addition to minimizing problems associated with feedstock variability during conversion, this capability could also assist in addressing sustainability issues by determining which material fractions are the most valuable as residues.

542a Life Cycle Assessment of Tallow-Based Renewable Diesel Fuel

Hong Jin, ConocoPhillips, 331 PL, Bartlesville Technology Center, HWY 60 & 123, Bartlesville, OK 74004 and Richard Nelson, Engineering Extension, College of Engineering, Kansas State University, 133 Ward Hall, Kansas State University,, Manhattan, KS 66506-2508.

Life cycle assessment (LCA) evaluates the impact of the utilization and fate of resources (feedstocks, process energy and efficiency, and waste disposal) on various commercial products from a systems analysis perspective. Recently the use of LCA for estimating the greenhouse gas (GHG) emissions for liquid transportation fuels has received increased attention. LCA is an important tool for evaluating renewable fuel technologies and their effects on GHG emissions. Several international, national, and state initiatives are formulating policies and regulations based on published LCA studies for fossil and biomass-derived fuels. However, the application of LCA in transportation fuel sector is still an evolving science. Some principal parameters and their implications, such as direct and indirect land use change, methodologies for treating non-fuel byproducts, and the underlying uncertainties in the input data and assumptions, are not well understood and deserve systematic investigation. The present work uses the LCA of tallow-based renewable diesel fuel as a case study to show the impact of such parameters.

542b Life Cycle Assessment of the Pyrolysis of Corn Stover into Green Gasoline and Green Diesel

David D. Hsu[1], Richard L. Bain[1], Richard E. Marinangeli[2], D.C. Elliott[3], and Scott Butner[3]. (1) National Bioenergy Center, National Renewable Energy Laboratory, 1617 Cole Blvd., MS 3322, Golden, CO 80401, (2) Renewable Energy and Chemicals Research, UOP LLC, Des Plaines, IL 60017, (3) Pacific Northwest National Laboratory, Richland, WA 99352

The pyrolysis of biomass, such as corn stover, can produce a bio-oil which can then be upgraded into diesel and gasoline using petrochemical refining technology. In addition to using a renewable feedstock, this pyrolysis process has the advantages of employing existing fuel refining, delivery, and end use infrastructure. A life cycle assessment (LCA) totaling environmental flows was performed on the corn stover pyrolysis process, from the production of the feedstock to the end use in a vehicle. The LCA used data from a corn stover E85 ethanol process[1], a National Renewable Energy Laboratory (NREL) wood pyrolysis AspenPlus model[2], as well as data and analysis from pyrolysis, hydrotreating, and hydrocracking runs by UOP LLC, Pacific Northwest National Laboratory, and NREL. The results from the LCA of the pyrolysis process will be compared to LCAs of a corn stover E85 ethanol process and a gasoline process. This LCA also shows that the production of corn stover is a particularly resource intensive step, with most of the greenhouse gas emissions resulting from this step. Additional life cycle inventory results will be presented, along with the assumptions and methodology used to complete the LCA.

[1]Sheehan, J.; Aden, A.; Riley, C. *Corn Stover Life Cycle Analysis*. NREL/FY01-220. Golden, CO: National Renewable Energy Laboratory, December 2000.

[2]Putsche, V. *Pyrolysis Process Design: Costs and Modeling*. NREL/FY04-545. Golden, CO: National Renewable Energy Laboratory, March 2004.

542c Life Cycle Analysis of Alternatives for Diesel: An Assessment of Inputs, Study Assumptions and Feedstock Type

Ken Koers[1], Tom N. Kalnes[2], and David R. Shonnard[1]. (1) Department of Chemical Engineering, Michigan Technological University, 1400 Townsend Drive, Houghton, MI 49931, (2) Research and Development, UOP LLC, 25 East Algonquin Road, Des Plaines, IL 60017

Background

With an increasing focus on renewable fuels and rising petroleum costs, it is vital to understand the environmental impacts from various alternative fuel products and processes. Biofuels are now seen as one of the simplest and most practical short-term fuel solutions. Substitutes for petroleum diesel are already widely popular in Europe, and gaining support in the US. Furthermore, current and future legislation aimed at assuring sustainability could result in regulation of biofuels by including greenhouse gas emission reduction and biodiversity requirements as determined by life cycle assessment (LCA). This further raises the need for a thorough grasp of the impacts of biofuels. Biodiesel and green diesel are the two most common diesel substitutes. Biodiesel is created through a transesterification reaction, producing a methyl ester which has similar properties to petroleum diesel. Green diesel is created through a catalyzed hydrogenation reaction on the plant oil, creating a liquid hydrocarbon fuel exhibiting superior performance for some properties compared to petroleum diesel. Syndiesel, another diesel fuel substitute, is created through high temperature and pressure gasification/catalytic reactions, from biomass, natural gas, or coal. To select from such a broad range of alternative diesel fuels will be challenging, but can be aided by a comprehensive environmental assessment.

Differences in study assumptions account for large variability in LCA impacts, making research into the effects of study assumptions important. The most important LCA study assumptions include methods of allocating inventory data among co-products, with or without farm emissions of N2O, land use change emissions, and feedstock type. To address these issues, this study characterized the greenhouse gas emissions, total energy, and fossil energy demand for biodiesel, green diesel, and syndiesel, and further analyzed the impacts of changing feedstock. The feedstock impacts are an ongoing activity in our laboratory, but plant oil feedstocks including algae, canola, jatropha, palm, rapeseed, soybeans, and sunflower will be reported on in this presentation. For syndiesel, feedstocks considered were wood residues, natural gas, and coal. Allocation methods based on product mass, energy content, and displacement were investigated in separate case studies.

Methods

Inventory data were obtained from several literature sources; from the US Department of Energy, from the European Joint Research Council of CONCAWE, from peer-reviewed publications, as well as biofuel processing data supplied by UOP LLC. As for the various life cycle stages, cultivation and feedstock processing data was obtained from literature sources covering a range of plant cultivation and geographic settings. SimaPro 7, a LCA software tool, providing data for some chemicals, electricity, heat, and other processes. Data for conversion from feedstock to fuel was provided by UOP LLC for biodiesel and green diesel, and for syndiesel, data came from government reports and publications. Data for N2O emissions from farm land due to fertilization was obtained from the various literature sources and also estimated using correlations from the most recent IPCC reports. Land use change is a complicated issue with very little agreement between studies. The clearing of land creates a "carbon debt", and literature sources give ranges from 5 years to 400 years for the time to pay this debt down when biofuels are produced and used. Our research will better understand this complex issue, and reduce the uncertainty of land use change.

Results

Preliminary results obtained to date in our laboratory show that biodiesel has exhibits a savings of between 48% and 65% in life cycle fossil energy demand over petroleum diesel, and savings of between 37% and 73% in greenhouse gas emissions. The ranges cited are a result of study assumptions with regard to inventory inputs, allocation method and feedstock type. Green diesel has life cycle fossil energy demand savings between 67% and 86% over petroleum diesel, and greenhouse gas savings between 42% and 85%. Syndiesel shows cumulative fossil energy demand ranging from 98% savings to a 51% increase over petroleum diesel, and greenhouse gas emissions range from a 91% reduction in emissions to a 115% increase, depending on fuel feedstock. Our preliminary results show green diesel having lower fossil energy and GHG impacts over the life cycle than biodiesel. Syndiesel results are very dependent on feedstock type, but can be either beneficial or detrimental. The effects of N2O could account for as much as 60% of the greenhouse gas footprint of a biofuel. Further investigation is underway to better understand the emissions of N2O from the field, a major contributor to the overall nitrous oxide emissions. Our continued research will help understand the reasons for these variations, and what effects are caused by the study assumptions chosen. Furthermore, the importance of input data is critical, and ensuring that the data comes from a reliable source. This study will bring all of these factors together for a comprehensive review on life cycle assessment, in the context of diesel fuel alternatives.

542d Developing a Sustainable Integrated Corn-Based Biorefinery

Robin E. Jenkins, Carina M. Alles, Susan M. Hennessey, Bruce M. Vrana, Jamie Ginn, Robert W. Sylvester, Julie Friend, and Mark H. Emptage. DuPont, Wilmington, DE 19898

Biofuels are part of a growing number of product and technology opportunities within the DuPont Applied BioSciencesTM Business. Cellulosic ethanol is one potential biofuel which provides a sustainable solution to the nation's growing concerns around energy supply and climate change. DuPont is leading the way towards demonstrating the feasibility of a cellulosic ethanol concept. DuPont received matching funds from the U.S. Department of Energy (DOE) for the Integrated Corn-Based Bio-Refinery (ICBR) program in collaboration with Diversa Corporation, the National Renewable Energy Laboratory (NREL), Michigan State University (MSU), and Deere & Company. DuPont scientists and engineers are committed to making advanced biofuels and energy-efficient biofuel processes a reality.

The ICBR program demonstrates how the early integration of sustainability analysis into the creative process of technology development enables a holistic approach to research guidance, where economic and environmental metrics are used alongside product performance standards to define and monitor success. Life Cycle Analysis, as a tool to evaluate the environmental profile of the ICBR value chain, has played a critical role in research guidance within the ICBR program as well as in science-based communications with external stakeholders.

542e Analysis and Design of Integrated Industrial and Ecological Systems for a Greener Cellulosic Ethanol

Anil Baral, Department of Chemical and Biomolecular Engineering, The Ohio State University, Columbus, OH 43210 and Bhavik R. Bakshi, Department of Chemical and Biomolecular Engineering, The Ohio State Unversity, Columbus, OH 43210.

Cellulosic ethanol is currently touted as a potential future sustainable transportation fuel that exerts lower environmental impacts, avoids food-fuel conflict, and more importantly improves energy security. Considering the nascent nature of cellulosic ethanol technology, it is imperative that its benefits and drawbacks be fully explored before the production begins in a commercial scale. Existing process–based life cycle studies provide favorable metrics for cellulosic ethanol in terms of energy and greenhouse gas emissions in comparison to corn ethanol and gasoline (1-5). However, the reliance of cellulosic ethanol production on ecosystem goods and services or natural capital is one important aspect that has not been rigorously examined. There are several critical resources such as fisheries, soil erosion, water, pollination, land, etc. which are as critical as fossil fuel consumption and need to be incorporated in decision-making. Ecologically-based lifecycle assessment (Eco-LCA) is an approach that accounts for contributions of ecological goods and services and helps track the natural capital footprint over the lifecycle (6). Such insight can also be used to design integrated industrial-ecological networks that reduce their impact.

A comprehensive study of cellulosic ethanol derived from five different feedstocks- switchgrass, corn stover, yellow poplar, recycled newspaper and municipal solid waste was conducted using hybrid Eco-LCA which integrates process-level lifecycle inventories with economic-scale life cycle inventories. A preliminary study shows that cellulosic ethanol E85 reduces greenhouse gas emissions by 38%-60% over gasoline. Energy returns on investment for cellulosic ethanol are higher with a range of 3-7. However, in terms of natural resource consumption cellulosic ethanol consumes more of all resources considered in the study except crude oil indicating its relatively higher resource intensity. Some examples of resources that are consumed more by cellulosic ethanol are natural gas, electricity, soil, crushed stone, water, various metal ores and nonmetallic ores. For corn stover-, yellow poplar- and switchgrass-derived ethanol, land use requirements are very large and it is constrained in the same way as corn ethanol. On the other hand, land use requirements for cellulosic ethanol derived from municipal solid waste and recycled newspaper are relatively small since these feedstocks do not require direct land use.

The Eco-LCA results are presented hierarchically with thermodynamic metrics based on energy and exergy used for aggregation. Overall, cellulosic ethanol consumes less non-renewable cumulative exergy which translates into better sustainability indices. Contribution of materials such as soil and detrital matter is significant implying that minimizing soil erosion, particularly for corn stover, would improve the overall system efficiency not including sunlight. Yield ratios of cellulosic ethanol from various feedstocks are better than that of corn ethanol but still lower than that of gasoline.

Our ongoing research work expands the earlier work by considering how industrial and ecological processes could be engineered to reduce the impact of the life cycle. For example, non-renewable inputs in the production system such as chemical fertilizers and fossil fuels may be replaced by biosolids, and renewable fuels by biodiesel and cellulosic ethanol. Such a model will be helpful in predicting reductions in natural resource/energy consumption and environmental impacts when biofuel production systems move closer to a sustainability ideal. The work will involve sensitivity analysis by varying biomass and ethanol yields. A scenario analysis that assumes the use of all lignocellulosic biomass available in the US for cellulosic ethanol production will be conducted. A theoretical scenario involving recycling of CO2 using a hydrogen fuel (7) will also be investigated. This approach has a potential for significantly minimizing land use requirements, and is appealing considering the constraint posed by limited available land for biomass production.

References:

(1) Sheehan, J.; Aden, A.; Paustian, K.; Kendrick, K.; Brenner, J.; Walsh, M.; Nelson, R. Energy and environmental aspects of using corn stover for fuel ethanol. J. Indust. Ecol. 2004, 7 (3- 4), 177-146.

(2) Spatari, S.; Zhang, Y.; MacLean, H.L. Life Cycle assessment of switchgrass- and corn stover-derived ethanol-fueled automobiles. Environ. Sci. Technol. 2005, 39 (24), 9750-9758.

(3) Kalogo, Y.; Habibi, S.; MacLean, H.L.; Joshi, S. Environmental implications of municipal solid waste-derived ethanol. Environ. Sci. Technol. 2007, 41 (1), 35-41.

(4) Levelton Engineering Ltd; S&T Consultants. Assessment of Net Emissions of Greenhouse Gases from Ethanol-Blended Gasolines in Canada: Lignocellulosic Feedstocks. Prepared for Agriculture and Agrifood Canada, Ottawa, ON, 1999, File 499-0893.

(5) Kemppainen, A. J.; Shonnard, D.R. Comparative life-cycle assessments for biomass-to-ethanol production from different regional feedstocks. Biotechnol. Prog. 2005, 21 (4), 1075-1084.

(6) Baral, A.; Bakshi, B.R. Hybrid LCA of corn ethanol, gasoline, soybean biodiesel, and diesel, part I – accounting for natural capital and net energy analysis. Environ. Sci. Technol., submitted 2008.

(7) Agrawal, R.; Singh, N. R.; Ribeiro, F. H.; Delgass, W. N. Sustainable fuel for the transportation sector. Proc. Natl. Acad. Sci. 2007 104 (12) 4828-4833.

542f Life Cycle Assessment of Renewable Transportation Fuel Feedstocks

Cynthia Folsom Murphy, Center for Energy and Environmental Resources, University of Texas at Austin, 10100 Burnet Rd, Blg 133, R7100, Austin, TX 78758, David Allen, Department of Chemical Engineering, University of Texas at Austin, Robert Hebner, Center for Electromechanics, University of Texas at Austin, and Michael Webber, Department of Mechanical Engineering, University of Texas at Austin.

Understanding the potential impact of large-scale production and use of alternative biofuel systems requires characterization and assessment of material and energy flows over the entire life cycle. "Well-to-wheel" analyses are being conducted for four potential, emerging renewable feedstocks suitable for production of transportation fuels: 1) ethanol from sugar cane; 2) ethanol from citrus waste; 3) biodiesel from cottonseed; and 4) biodiesel from algae. These biofuels are compared to four existing fuel systems: gasoline, corn ethanol blends (E85), soybean biodiesel, and petroleum diesel. The framework used to evaluate these feedstocks and fuels is a set of process-flow models that account for inputs and outputs and which are designed for economic and environmental assessment and comparison. The process flow models, developed for all eight feedstocks, are designed to be dyanamic and modular; the inputs are thus parametric and the modules are developed at the unit operation level. In combination, the modules cover the full life cycle of the various fuels systems from raw material acquisition through vehicle combustion, with appropriate allocation of by-products. Detailed, quantitative, activity-based models are designed to address agricultural activities, transportation to processing facilities, conversion of feedstocks into fuels for powering passenger vehicles, and fuel delivery systems. Environmental metrics considered in the study include net energy, water use, land use, and emissions to air (both greenhouse gases and criteria pollutants). The geographic boundaries are the U.S., with the exception of petroleum feedstock production, in which case both domestic and foreign supplies are included. Assessment of the environmental impact of fuel combustion (i.e., the use phase) will be performed separately by modeling emission profiles and by incorporating the results into photochemical air quality modeling simulations to estimate regional air quality impacts. The final outcome of this project is expected to be an environmental characterization of emerging renewable feedstocks and fuels, along with an evaluation of potential economic and technological barriers to widespread adoption. Results will be presented in such a manner as to be understood by the general public, thus facilitating more informed choices by both consumers and policymakers. Furthermore, the flexible, dynamic nature of the models will facilitate evaluation of additional fuels systems of interest.

573a Pretreatment of Rice Straw with Ammonia Recycled Percolation

Choon-Hyoung Kang, Chemical Engineering Department, Chonnam National University, 300 Yongbong-dong, Buk-ku, Kwangju, South Korea and Y. Y. Lee, Chemical Engineering, Auburn University, Auburn, AL 36849.

Because of high contents of cellulose (~37 wt%) and hemicellulose (~17%), rice straw seems to be a potential lignocellulosic biomass for production of bioethanol. In this study, Ammonia Recycled Percolation (ARP) pretreatment of rice straw was extensively investigated. In particular, the experimental study included the effects of temperature, reaction time and concentration of ammonia on compositions and enzymatic digestibility of the resulting solid residues; the ranges of pretreatment conditions were, in turn, 150°C~210°C, 10 min ~ 90 min and 0~ 20 wt%. Through ARP pretreatment, the lignin content was reduced by as high as ~84% while 20~80 % of the hemicellulose was solubilized. The solid residue resulted from the pretreatment with 15 wt% aqueous ammonia solution at 170°C for 90 mim showed as high as ~90% of digestibility with enzyme loading of 15FPU/g of glucan. Addition of the xylanese along with cellulase led to a notable enhancement of digestibility, indicating a possible inhibitory role of hemicellulose. Simultaneous Saccharification and Fermentation (SSF) and Simultaneous Saccharification and Co-Fermentation (SSCF) were performed to obtain ethanol productions of 1.38 g/L (corresponding to 81% yield) and 15 g/L (corresponding to 89% yield), respectively.

573b Effects of Dilute Acid Hydrolysis Conditions on Total Sugar Yields from Enzymatic Hydrolysis of North Central Forest Feedstocks

Michael J. Brodeur-Campbell, Jill I. Jensen, Juan Morinelly, and David R. Shonnard. Chemical Engineering, Michigan Technological University, 1400 Townsend Dr., Houghton, MI 49931

Motivation

Due to the combination of increasing demand for oil, decreasing supply, and greenhouse gas (GHG) emissions, biofuels are increasingly being investigated as carbon neutral alternatives to petroleum fuels. Current ethanol production from corn starch is nearing the feasible maximum, and only delivers approximately 1.3 units of energy in the fuel product for every unit of fossil energy consumed in the production (1). Producing fuel from starch has the additional negative effect of competing directly with the food supply and has potential for GHG emissions from large-scale land use change. Cellulose is the most abundant biopolymer on Earth, and lignocellulosic feedstocks have the potential to provide greater than 10 units of energy in the fuel for each unit of fossil energy consumed in the production (2). However, the technology for converting cellulose to sugar is still challenged with technical barriers and much work is needed to bring the cost of production down to the point where it is economically competitive.

Michigan's Upper Peninsula comprises an area approximately 14,500 square miles situated in the north central forested region of the United States. Due to ecological constraints on the growing season, agriculture is significantly limited. However the region has historically been and remains a productive forest ecosystem which supplies much timber for construction and pulp for paper, and has the potential to provide a significant amount of lignocellulosic biomass to future biorefineries. Furthermore, abandoned agricultural lands are in abundance that could be converted to energy crop production.

The goal of this research is to determine the effects of dilute acid hydrolysis pretreatment conditions on the maximum total sugar yield (pentoses plus hexoses) after pretreatment and enzymatic hydrolysis by commercially available cellulase mixtures. An understanding of the effect one process has on subsequent processes downstream will be important to future industrial operations. The data will provide yield information on a number of important forest species and lay a foundation for future enzyme improvement studies.

Methods

Dilute acid hydrolysis is a common method of pretreating lignocellulosic feedstocks prior to enzymatic hydrolysis. Dilute acid hydrolysis removes the hemicellulose fraction – composed primarily of five-carbon or pentose sugars – leaving behind cellulose and lignin which are much more recalcitrant to hydrolysis. Enzymatic hydrolysis frequently follows pretreatment to liberate hexose sugars – primarily glucose – from cellulose, leaving behind only lignin which can be used as a boiler fuel to provide heat and electricity for the process.

The primary goal of pretreatment is to expose the cellulose for more efficient enzymatic hydrolysis. However two competing secondary considerations are also important in the overall process; pentose sugars from dilute acid hydrolysis are fermentable by some bacteria species thereby increasing the total potential ethanol yield, but degradation products from dilute acid hydrolysis – primarily furfural – decrease total sugar yields and inhibit fermentation.

Three species of locally available and commercially relevant biomass feedstocks will be considered; aspen (a hardwood), balsam (softwood), and switchgrass (model energy crop). Dilute acid hydrolysis (pretreatment) conditions to be varied include temperature (150°C to 175°C), sulfuric acid concentration (0.25%, 0.50%, and 1.00%), and biomass species (aspen, balsam, and switchgrass). Pretreatment reactions will occur at constant temperature using ten 5 ml liquid volume stainless steel tubular reactors immersed in 750 ml of a Dow Corning 550 heat transfer fluid in a Parr Instruments Model 4570 reactor. Reactor tubes will be removed at different times from the hot oil bath and placed in ice to stop the reaction. Samples of the hydrolysate will be taken from the tubular reactors, filtered (0.22 μm pore size) and analyzed by High Performance Liquid Chromatography (HPLC) for sugar content. Solid residues will be washed with distilled water and enzymatic hydrolysis will be performed on the highest sugar yielding tubular reactor for each set of pretreatment conditions. Enzymatic hydrolysis conditions will be: 1% dry weight pretreated biomass (cellulose plus lignin), 60 filter paper units (FPU) SpeezymeCP (Genencor) per dry gram cellulose plus 120 cellobiohydrolase units (CBU) beta-glucosidase (Novozyme, Novozym 188) per dry gram cellulose, 72 hr. reaction time, and 50°C.

Results

Preliminary results using a similar "mini-reactor – 2.5 ml" pretreatment experimental system exhibited a distinct effect of pretreatment conditions on xylan sugar yields from pretreatment and glucan sugar yields from enzymatic hydrolysis. More severe pretreatment reduced xylan sugar yields and increased byproduct concentrations (furfural), but also increased glucose yields from enzymatic hydrolysis, especially at early in the 72 hr hydrolysis period for hardwoods (aspen, red maple, basswood) and also at all times for balsam (softwood). Results from the proposed design of experiments will add to those already obtained to identify optimum conditions for extracting maximum total sugars from the hydrolysis of the forest biomass species included in this study. Recommendations for dilute acid hydrolysis conditions to maximize total fermentable sugar yield will be made.

References

(1) USDA. 2002. The Energy Balance of Corn Ethanol: An Update, AER-814, U.S. Department of Agriculture Office of the Chief Economist and Office of Energy Policy and New Uses.

(2) U.S. DOE. 2006. Breaking the Biological Barriers to Cellulosic Ethanol: A Joint Research Agenda, DOE/SC-0095, U.S. Department of Energy Office of Science and Office of Energy Efficiency and Renewable Energy (www.doegenomestolife.org/biofuels/).

573c Kinetic Characterization of Oligomer Formation and Degradation in Dilute Acid Pretreatment of Lignocellulosic Biomass from Forests

Juan Morinelly[1], Jill R. Jensen[1], and David R. Shonnard[2]. (1) Chemical Engineering, Michigan Technological University, 1400 Townsend Dr., Houghton, MI 49931, (2) Department of Chemical Engineering, Michigan Technological University, 1400 Townsend Dr., Houghton, MI 49931

Introduction

Dilute acid pretreatment achieves the saccharification of the hemicellulose portion of lignocellulosic biomass in a biomass to ethanol process. The yield of fermentable, monomer sugars during pretreatment is a crucial factor for the maximization of the overall process efficiency.

The kinetics of xylose formation and degradation has been investigated in more detail than that of any other sugar present in hemicellulose. This is the case given that xylose is often the main component with percentages of total hemicellulose by dry weight ranging from 27 to 91% depending on the type of feedstock [1]. The simplest kinetic mechanisms that has been proposed for this reaction is a two-step pseudo-first order irreversible reaction with Arrhenius-type kinetic constants [2]:

$$Hemicellulose\,(H) \xrightarrow{k_1} Xylose\,(X) \xrightarrow{k_2} Degradation\ products\,(D)$$

However, experimental observations suggest that other models that include oligomeric intermediates and parallel reactions of slow and fast reacting hemicellulose phases (biphasic) could describe the reaction more accurately [3]. There is not a consensus on whether the biphasic behavior of hemicellulose is real or simply an artifact of experimental methods. On the other hand, oligomer concentrations have been measured [4-6] indicating that an oligomer intermediate should be included in kinetic modeling:

$$Hemicellulose\,(H) \xrightarrow{k_1} Oligomer\,(O) \xrightarrow{k_1'} Xylose\,(X) \xrightarrow{k_2} Degradation\ products\,(D)$$

The need to understand oligomer kinetics is further justified because oligomer sugars are not as fermentable as monomer sugars. The objective of this research is to develop an experimental procedure in order to determine kinetic parameters that can be fitted to the oligomeric intermediate model above for hardwood, softwood, and herbaceous species (Aspen, Balsam, and Switchgrass respectively). The kinetic parameters will be used in follow-on research to predict the effects of dilute acid hydrolysis conditions on the dynamics of oligomer formation and degradation and modify reactor design to achieve minimum oligomer and maximum monomer sugar concentrations.

Methods

Previous experiments in our lab using a transient temperature method showed significant oligomer sugar formation early in the reaction time for xylose, glucose, mannose, galactose, and arabinose. As time in the reaction increases, oligomer concentration decreases as they are hydrolyzed to monomer sugars. The research to be reported here extends this early work using isothermal small-scale tubular reactors (3/8" OD, 0.035" wall, ~6 mL working volume).

A constant temperature bath at a temperature between 150 and 200 °C is achieved using a suitable silicon oil, *Dow Corning 550®*. Once thermal equilibrium has been achieved, a basket with 10 stainless steel tubular reactors containing the dilute acid solution and the biomass (.25, 0.50, and 1% H_2SO_4, temperature of 150°C, 160°C and 175°C, 10 % biomass solids, biomass species aspen, balsam, switchgrass) is submerged in the bath. This design of experiments will cover a broad range of key reaction parameters to characterize the kinetics of hydrolysis and identify conditions for minimum oligomer sugar concentrations.

According to predictions made using heat transfer simulation software, *COMSOL Multiphysics*™ the time necessary for the contents of the tubular reactor to reach the target temperature by conduction is on the order of 4 minutes. Additionally, an analytical solution to a simplified limiting case (well mixed reactors) predicts a heat transfer period on the order of 10 seconds. In order to obtain composition data at isothermal conditions this heat transfer time must be kept short by mixing.

The tubular reactors are retrieved at different times from the silicon oil bath and immediately placed in an ice bath in order to stop the reaction. The first sample is taken at a pre-established heat transfer time, and the measured compositions for that sample will be the initial conditions. This new set up allows retrieving the remaining 9 tubular reactors with high time resolution, which is increasingly important as the experiments are undertaken at the higher end of the desired temperature range and the reaction times decrease exponentially.

The hydrolyzate from each tubular reactor sample will be analyzed for sugar concentrations via an HPLC, and oligomer concentrations are to be determined by a secondary hydrolysis in the same fashion as in previous studies from our lab [5].

Once the kinetic parameters for the three step kinetic model are determined for the mentioned species, the prediction for concentration profiles at different temperatures and acid concentrations will be compared to experimental data and the validity of the model and its predictions will be determined.

References

1. McMillan, J.D., *Process for Pretreating Lignocellulosic Biomass: A Review. NREL/TP-421-4978.* 1992.

2. Saeman, J.F., *Kinetics of Wood Sacharification: hydrolysis of Cellulose and Decomposition of Sugars in Dilute Acid at High Temperature.* Industrial & Engineering Chemistry Research, 1945. **37**(1): p. 43-52.

3. Jacobsen, S.E. and C.E. Wyman, *Cellulose and Hemicellulose Hydrolysis Models for Application to Current and Novel Pretreatment Process.* Applied Biochemistry and Biotechnology, 2000. **84-86**: p. 81-96.

4. Yat, S.C., A. Berger, and D.R. Shonnard, (2008) *Kinetic Characterization for Dilute Sulfuric Acid Hydrolysis of Timber Varieties and Switchgrass.* 99(9), pp 3855-3863.

5. Jensen, J.R., Morinelly, J., Aglan, A., Mix, A., Shonnard, D.R., Kinetic Characterization of Biomass Dilute Sulfuric Acid Hydrolysis: Mixtures of Hardwoods, Softwood, and Switchgrass, *AIChE Journal*, online April, 2008, http://www3.interscience.wiley.com/journal/109931315/issue

6. Chen, R., Y.Y. Lee, and R. Torget, *Kinetic and Modeling Investigation on Two-Stage Reverse-flow Reactor as Applied to Dilute-Acid Pretreatment of Agricultural Residues.* Applied Biochemistry and Biotechnology, 1996. **57/58**: p. 133-146.

573d Selection for Improved Hybrid Poplar Via Dilute Acid and Enzymatic Hydrolysis Using Mini-Reactors

Jill R. Jensen[1], Anwar Aglan[1], Victor Busov[2], and David R. Shonnard[3]. (1) Chemical Engineering, Michigan Technological University, 1400 Townsend Dr., Houghton, MI 49931, (2) School of Forest Resources and Environmental Science, Michigan Technological University, 1400 Townsend Dr., Houghton, MI 49931, (3) Department of Chemical Engineering, Michigan Technological University, 1400 Townsend Dr., Houghton, MI 49931

Large-scale production of transportation fuels from woody biomass will result in expansion of land dedicated to biomass production. Improved hybrid poplar, along with other high-yielding energy crops such as switchgrass and hybrid willow, have the potential to supply feedstocks to an emerging biofuels industry with lower land footprints than wild undomesticated species. Lignocellulosic ethanol, derived from woody biomass, is an alternative to conventional gasoline that can be produced from fermentation of sugars derived from woody biomass. One of the preferred methods for the production of fermentable sugars from biomass is a dilute acid hydrolysis pretreatment followed by an enzymatic hydrolysis. The goal of these processes is to obtain the highest conversion possible, thus leading to the greatest ethanol yield. A venue for increasing ethanol yields is to genetically modify the biomass to either increase total sugar content or alter the structure in a way to make the sugars more accessible under less aggressive pretreatment conditions. Genetic modifications targeted at improving biomass properties involve testing large amount of progenies and/or genetically modified transgenics. This requires high throughput and a small amount of sampling material. The results presented here demonstrate "mini-reactors" for dilute acid and enzymatic hydrolysis which can be set up for medium to high throughput studies.

The experiments were conducted with hybrid poplars modified for rate and extent of secondary growth. Eight different samples were studied including three wild-type (Control 3C, Control 4D, and Control 5E) and five transgenic poplars (BL 121 (1), BL 121 (2), BL 12, BL 161, and BL 204). The goal of the experiments was to determine whether the genetic modifications will result in higher total sugar yields from dilute acid hydrolysis pretreatment followed by enzymatic hydrolysis and test the feasibility using the mini-reactor for these studies. A reduction in size of 200x for each experiment was necessary to accommodate the extremely small size of the sample (less than 5 grams total for each hybrid poplar sample).

Two mini-reactors were used; each consisted of a stainless steel Swagelok Fitting with a ½ inch tube outer diameter and two Stainless Steel Swagelok Plugs; and a volume of approximately 3 ml. 0.12g solids were added to each reactor with 2.4 ml of 0.5% sulfuric acid. The optimum maximum temperature for these experiments was determined to be 170 degrees Celsius. The mini-reactors were secured to the cooling coils on the top of a 1-L well-mixed batch reactor (Parr Instruments, Model 4570 reactor) using stainless steel wire. The bottom of the Parr reactor was filled with approximately 900 ml of distilled water and the top was secured in place. Reactor temperature was increased from room temperature to 170 degrees Celsius at a rate of approximately 3 degrees C/min, and upon reaching the maximum temperature the reactor was cooled at a rate of 10 degrees C/min using tap water. One sample of the hydrosylate was taken from each reactor at the end of the experiment and analyzed by High Performance Liquid Chromatography (HPLC) for sugar content.

The hydrosylated biomass was removed from the mini-reactors, washed, and subjected to enzymatic hydrolysis. This procedure was also scaled back, more than 100x compared to the procedure previously used. Enzyme loadings of 60FPU Spezyme CP (Genencor) / dry gram substrate and 2 CBU/FPU of beta-glucosidase (Novozyme, Novozym 188) were used. Enzymatically digested samples were taken at the end of 72 hours and analyzed by HPLC for sugar content.

Results from the three control samples indicate that the small scale pretreatment experiments were highly reproducible with respect to xylose yield and yields of the minor sugars, both from trial to trial in the mini-reactors, and also compared to larger volume trials at the 0.5 L scale. BL204 transgenic modification showed enhanced xylose yield compared to the controls, whereas the other transgenics exhibited smaller xylose yields. The overall yield of total sugars (glucose, xylose, galactose, arabinose, and mannose) generated by both the pretreatment and enzymatic hydrolysis expressed as percent of the weight of the starting raw material ranged from 52-62%. Two of the genetically-modified poplars, BL 161 and BL 204, appear to have statistically significant higher total sugar yields (61-62%) compared to controls and other hybrid poplar samples (e.g., yields ranging from 52-56%). We are currently investigating the structural and chemical properties of the modified materials to better understand the changes resulting in the observed higher sugar yields.

The small scale pretreatment and enzymatic hydrolysis experimental procedures developed for this work allow for reproducible analysis of samples on a much smaller scale. With the experimental setup previously used, it could take years for enough wood biomass to be produced for testing the effect of the genetic modifications. Here we demonstrate a device and procedure that achieves significant reduction (less than five grams) in sample size needed. This coupled with reproducibility of the results allows for more convenient, intensive and high throughput studies of lignocellulosic conversion properties.

573e Effects of Reactor Type on High Solids Enzymatic Hydrolysis of Dilute Acid Pretreated Corn Stover

Christine M. Roche, Jonathan J. Stickel, and Richard T. Elander. National Bioenergy Center, National Renewable Energy Laboratory, 1617 Cole Blvd, Golden, CO 80401

Ethanol produced from lignocellulosic biomass has become a promising alternative transportation fuel to gasoline. In a typical biochemical process, the majority of the hemicellulose fraction of biomass is hydrolyzed in a chemical pretreatment process, while the majority of the cellulose fraction is converted through enzymatic hydrolysis. A combination of pretreatment and enzymatic hydrolysis performed at high concentrations of solids can potentially improve process economics. At high concentrations of solids, smaller reactors are needed for the same product throughput, which decreases capital costs and increases product concentration. However, at high solid levels there can be difficulties in pumping, mixing, pH and temperature control, and lower total conversion of cellulose. In this work different scaled-down enzymatic hydrolysis reactors were tested and evaluated for their efficiency in converting high solids loadings of biomass. Three types of reactors were compared: vertical stirred tank, horizontal tank with center drive shaft, and horizontally rotating jars. In addition, the horizontally rotating jars were varied in number of baffles and amount of attrition media. The affects of mixing speed, presence of baffles, and amount of attrition media are shown. Cellulose conversion results are compared for all reactors.

573f Equilibrium Adsorption Characteristics of Polysaccharide Adsorbents

Youngmi Kim[1], Rick Hendrickson[1], Nathan S. Mosier[2], and Michael R. Ladisch[1]. (1) Laboratory of Renewable Resources Engineering (LORRE), Purdue University, Potter Engineering Center, 500 Central Drive, West Lafayette, IN 47907-2022, (2) Agricultural and Biological Engineering & Laboratory of Renewable Resources Engineering (LORRE), Purdue University, Potter Engineering Center, 500 Central Drive, West Lafayette, IN 47907-2022

Starch based adsorbents have found utility in the fuel ethanol industry to remove water from fuel ethanol in an energy efficient manner. Corn grits packed in a fixed bed absorber are used in a thermal swing mode to remove water from sub-azeotropic ethanol to yield a fuel-grade alcohol product. The heat of adsorption is stored in the bed, and combined with a counter-flow of CO_2 (a co-product of the ethanol fermentation), this heat provides the energy required to regenerate the bed. A multi-bed, staged operation is used to obtain a continuous stream of fuel-grade product ethanol. A challenge in modeling this system has been the paucity of equilibrium data, which is obtained at representative conditions. A dynamic method has been developed that enables equilibrium data to be obtained in a flow-through system. Equilibrium adsorption of water on starch based adsorbents has been determined for sub-azeotropic ethanol that contains 90-95% (v/v) ethanol. Linear isotherm is valid for this range of feed ethanol. The adsorbent is packed in a basket inserted in a long column. Flow through the basket is managed to minimize dispersion, and enable the extent of adsorption to be measured using the change in composition between inlet and outlet hydrous ethanol streams as well as regenerant condensate composition. Equilibria data, a model, and the method are presented in this paper. These data are used to model the dynamic operation of a fixed bed system where the feedstock for corn ethanol is also the separations media.

573g Fractionating Recalcitrant Miscanthus by a Two-Stage Treatment Under Mild Reaction Conditions

Bin Wang[1], Xiaojuan Wang[2], and Hao Feng[1]. (1) Energy Biosciences Institute and Department of Food Science & Human Nutrition, University of Illinois at Urbana-Champaign, 382D-AESB, 1304 W Pennsylvania Ave, Urbana, IL 61801, (2) Department of Food Science and Human Nutrition, University of Illinois at Urbana-Champaign and Dalian University of Technology, 174-AESB, 1304 W Pennsylvania Ave, Urbana, IL 61801

Overcoming the natural recalcitrance of biomass to microbial breakdown is essential to any operation that converts biomass to fermentable sugars. Pretreatments provide the key to unlocking the protective structures in biological feedstocks so that enzymatic hydrolysis of the carbohydrate fraction to monomeric sugars can be achieved more rapidly and with greater yield. Numerous pretreatment concepts have been proposed and tested over the years. Steam treatment (or steam explosion) with and without the addition of a catalyst is one of the oldest methods. A coordinated study supported by a USDA Initiative for Future Agricultural and Food Systems (IFAFS) grant has examined, in phase I of the project, the performance of five promising biomass pretreatment methods. Almost all the five pretreatment methods, i.e., dilute acid, hot water (neutral pH), ammonia fiber/freeze explosion (AFEX), ammonia recycle percolation (ARP), and lime can produce high sugar yield. Nevertheless, it is found that low cost pretreatment reactors are often counterbalanced by the higher costs associated with either pretreatment catalyst recovery or higher costs for ethanol product recovery. Although a number of technically feasible approaches for biofuel production from lignocellulosic biomass have been explored, the development of economically feasible biofuel production methods remains a key challenge. Up to 67% of the total cost in cellulosic ethanol production is the processing costs and pretreatment is the most expensive single unit operation. While researchers are scrutinizing each step in biomass conversion to minimize the production costs, the concept of biorefinery puts effort on enhancing the cost structure of biofuel generation. Similar to its successful counterpart in the petrochemical industry, the ultimate goal of a biorefinery is the efficient fractionation of lignocellulosic biomass into multiple streams that contain value-added compounds so that the overall economics of a biofuel production facility can be significantly improved. A two-stage concept was used in this study to fractionate Miscanthus into valued-added products (hemicelluloses and lignin) and fermentable sugars under relatively mild condition. Using catalysts such as acidic and alkaline electrolyzed water, dilute sulfuric acid, or alkaline hydrogen peroxide, the digestibility of Miscanthus by a one-step pretreatment at 160, 170, 180, 190, 200, or 210 oC for 8, 16, or 24 min was compared with a two-stage fractionation at 50 and 121 oC. FTIR was used to characterize structure changes in the pretreated samples.

573h Enhancing CO-Water Mass Transfer with MCM41 Nanoparticles and Electrolytes

Haiyang Zhu, Chemical and Biological Engineering; Mechanical Engineering, Iowa State University, 2025 Black Engineering; 2114 Sweeney Hall, Iowa State University, Ames, IA 50011, Brent H. Shanks, Department of Chemical and Biological Engineering, Iowa State University, 2119 Sweeney Hall, Ames, IA 50011, and Theodore J. Heindel, Department of Mechanical Engineering, Iowa State University, 2025 Black Engineering Building, Ames, IA 50011.

Synthesis gas (syngas) fermentation is a potential pathway for converting biomass to a variety of fuels and chemicals such as ethanol, methane, and acetic acid due to its low cost and potentially high efficiency.[1] In syngas fermentation, carbon monoxide is the sole carbon source from which the microorganisms produce their desired chemicals. However, the CO solubility in the aqueous media is low, which minimize the CO-liquid mass transfer rate and limits the overall product yield.[2] For syngas fermentation to be successful, the CO-liquid mass transfer rates must be improved. However, limited attention has been paid to unique approaches for enhancing CO-liquid volumetric mass transfer coefficients. In general, gas-liquid mass transfer is influenced by reactor type, power input, and system additives. Olle et al.[3] reported the oxygen-liquid mass transfer could be enhanced by up to 600% in an agitated sparged reactor when magnetite (Fe_3O_4) nanoparticles coated with oleic acid and a surfactant were added to the systems. Others have also shown that gas-liquid mass transfer rate can be enhanced by adding micro- or nanoparticles to the system.[4-6] In addition, the introduction of electrolytes can inhibit bubble coalescence, thereby increasing the gas-liquid interfacial area.[7]

As reviewed by Beenackers and van Swaaij,[8] the enhancement in gas-liquid mass transfer by adding particles was influenced greatly by particle size and surface characteristics (e.g., hydrophobic or hydrophilic). Mesoporous silica materials (MCM41) have extensive adsorption capacity due to their large BET surface area (~1000 m2/g), which can also be readily functionalized with organic groups to adjust the surface characteristics such as the hydrophobicity or hydrophilicity and acidity or basicity.[9-11] Hence, this material is a good candidate for enhancing the CO-liquid mass transfer rate.

MCM41 nanoparticles were prepared following the procedures of Deng et al.[12] TEM images showed MCM41 nanoparticles had a spherical morphology with a diameter of ~250 nm, which is supported by particle size distribution results. The BET surface area of MCM41 nanoparticles is ~1200 m2/g. Various organic groups such as methyl (Me), carboxylpropyl (CP), aminepropyl (AP), mercaptopropyl (MP), and mercaptoundecyl (MU) are grafted to the MCM41 nanoparticles to adjust their surface characteristics. After grafting Me groups, functionalized nanoparticles (MeMCM41) had the similar morphology and size as the MCM41 nanoparticles and the BET surface area was ~1300 m2/g.

Various electrolytes solutions were also investigated. Manganese (II) sulfate monohydrate and cobalt (II) sulfate heptahydrate were purchased from Sigma-Aldrich; cobalt (II) nitrate was purchased from Acros organics; manganese (II) chloride tetrahydrate, nickel (II) sulfate hexahydrate, nickel (II) nitrate hexahydrate, nickel chloride hexahydrate, cupric (II) nitrate, magnesium sulfate heptahydrate, ferrous (II) sulfate heptahydrate, and copper (II) sulfate pentahydrate were purchased from Fisher scientific.

The CO-water volumetric mass transfer enhancement results showed that silica particle size and preparation procedure greatly affected the enhancement. When the surfactant involved in the MCM41 synthesis was removed by extraction, the enhancement increased from 1 to 1.55 as the nanoparticles concentration increased from 0 to 0.4 wt%. For large silica particles with diameters of 1.4 and 7 μm, the maximum enhancement was 1.29 and 1.01, respectively. In addition, MCM41 nanoparticles after calcinations at 500oC showed a lower enhancement of ~1.11. These results suggested the particles size affected the CO-water mass transfer rate and surface hydroxyl groups played an important role in the CO-water mass transfer enhancement.

Mercaptopropyl or mercaptoundecyl groups functionalized MCM41 nanoparticles with the functional group loading of 5% (molar ratio) showed a continuous increase in enhancement with increase in nanoparticle concentration, with the highest enhancement of 1.9 when the nanoparticle concentration was 0.4 wt%, which is larger than those samples with other functional groups. In addition to hydrophobicity differences, the mercaptan groups on the MPMCM41 and MUMCM41 naoparticle surface seemed to contribute to the adsorption of CO. In-situ CO adsorption FT-IR spectra indicated the adsorbed CO species on MPMCM41 nanoparticles was more stable than that on MCM41 and NPMCM41 nanoparticles. Hence, it is hypothesized that when mercaptopropyl or mercaptoundecyl groups were grafted to the MCM41 nanoparticles, the nanoparticles adsorbed more CO molecules upon contacting the CO bubbles and they released CO molecules into the water, which resulted in the extra enhancement of the CO-water volumetric mass transfer coefficients compared to pure MCM41 nanoparticles. The enhancement in CO-water mass transfer coefficient (kL) also indicated the addition of MCM41 nanoparticles both increased the CO-water interfacial area (a) and CO-water mass transfer coefficient (kL).

When different electrolytes were added to the system at the concentration of 5 wt%, the CO-water mass transfer was enhanced by 1.6~4.7, depending on the electrolyte type, with copper sulfate providing the strongest enhancement of 4.7. For those electrolytes with the same cations, sulfate electrolytes showed better enhancement than other cations such as chloride and nitrate electrolytes. In addition, it is believed that the anion type gave stronger effect than the cation type. The enhancement resulted from the inhibition of bubble coalescence by electrolytes. In addition, the water-air interface had a preferential orientation, with the oxygen atoms outermost.[13] As a result of this orientation, an electrical double layer was established at the surface, with the outermost portion of the double layer being negative and the innermost part being positive. With the positive portion of the double layer directed into the solution, the anions preferentially accumulated near the interface, leading to a stronger effect by anions on the CO-water mass transfer enhancement compared to cations.

The CO-water mass transfer coefficient enhancement (Ek) results suggested the enhancement to the CO-water mass transfer rate by adding nanoparticles or electrolytes is attributed to one of two reasons. First, the addition of nanoparticles enhanced both the CO-water mass transfer coefficient (kL) and the CO-water interfacial area (a). Second, electrolyte addition only enhanced the CO-water interfacial area by suppressing bubble coalescence.

References:

Acknowledgement:

This material is based upon work supported by the Board of Regents, State of Iowa, through the Battelle Infrastructure and Platform Grants Program.

(1) Do, Y. S.; Smeenk, J.; Broer, K. M.; Kisting, C. J.; Brown, R.; Heindel, T. J.; Bobik, T. A.; DiSpirito, A. A. Biotechnology and Bioengineering 2007, 97, 279-286.

(2) Lynd, L. R. Annual Review of Energy and the Environment 1996, 21, 403-465.

(3) Olle, B.; Bucak, S.; Holmes, T. C.; Bromberg, L.; Hatton, T. A.; Wang, D. I. C. Industrial & Engineering Chemistry Research 2006, 45, 4355-4363.

(4) Py, X.; Roizard, C.; Bergault, I.; Midoux, N. Chemical Engineering Research & Design 1995, 73, 253-257.

(5) Littlejohns, J. V.; Daugulis, A. J. Chemical Engineering Journal 2007, 129, 67-74.

(6) Dagaonkar, M. V.; Beenackers, A. A. C. M.; Pangarkar, V. G. Chemical Engineering Journal 2001, 81, 203-212.

(7) Zahradnik, J.; Fialova, M.; Kastanek, F.; Green, K. D.; Thomas, N. H. Chemical Engineering Research & Design 1995, 73, 341-346.

(8) Beenackers, A. A. C. M.; Vanswaaij, W. P. M. Chemical Engineering Science 1993, 48, 3109-3139.

(9) Schumacher, C.; Gonzalez, J.; Perez-Mendoza, M.; Wright, P. A.; Seaton, N. A. Recent Advances in the Science and Technology of Zeolites and Related Materials, Pts a - C 2004, 154, 386-393.

(10) Schumacher, C.; Gonzalez, J.; Perez-Mendoza, M.; Wright, P. A.; Seaton, N. A. Industrial & Engineering Chemistry Research 2006, 45, 5586-5597.

(11) Munoz, B.; Ramila, A.; Perez-Pariente, J.; Diaz, I.; Vallet-Regi, M. Chemistry of Materials 2003, 15, 500-503.

(12) Deng, W. H.; Toepke, M. W.; Shanks, B. H. Advanced Functional Materials 2003, 13, 61-65.

(13) Jarvis, N. L.; Scheiman, M. A. Journal of Physical Chemistry 1968, 72, 74-78.

573i Use of Immunolabeling to Study Changes in the Distribution of Xylan in the Plant Cell Wall to Better Understand the Process of Xylan Hydrolysis during Dilute Acid Pretreatment

David K. Johnson, Roman Brunecky, Ashutosh Mittal, Todd B. Vinzant, and Michael E. Himmel. Chemical and Biosciences Center, National Renewable Energy Laboratory, 1617 Cole Boulevard, Golden, CO 80401

A key barrier to the commercialization of fuels and chemicals produced from lignocellulosic biomass is the high cost and relative inefficiency of converting biomass into fermentable sugars. Dilute acid pretreatment is a promising technology for increasing the enzymatic digestibility of lignocellulosic biomass. A better understanding of the pretreatability of biomass is needed so that the rate of formation and yields of sugars can be increased. Xylan is an important hemicellulosic component of the plant cell wall and acts as a barrier to cellulase access to cellulose. A key function of the dilute acid pretreatment process is the hydrolysis and solubilization of xylan so that cell wall sugars are available for subsequent fermentation. To better understand xylan hydrolysis in corn stover we have studied changes in the distribution of xylan caused by dilute acid pretreatment.

Samples of corn stover rind were labeled with a xylan antibody (LM11) that was conjugated with a fluorescent dye (Alexa Fluor488). Changes in the distribution of xylan were then monitored using laser scanning confocal microscopy. Samples were also labeled with another xylan antibody (LM10) that was conjugated with a gold nanoparticle. After subsequent gold enhancement, xylan distribution within the cell wall could be visualized by scanning electron microscopy. Our studies have focused on the sclerenchyma cells near the vascular bundles for consistency in imaging and because these cells also contain the largest amounts of mass. After pretreatment the xylan antibody signal was observed to decrease in the bulk of the primary and secondary cell walls while the signal in the middle lamella and lumen was maintained and possibly even increased. It is possible that these changes in xylan distribution may be related to the biphasic kinetic behavior that has been reported for xylan hydrolysis rates.

573j Improving the Quality of Corn Stover Biomass through Biological Pretreatment

Deepti Tanjore[1], Tom L. Richard[2], and Megan Marshall[2]. (1) Agricultural and Biological Engineering, Pennsylvania State University, 249, Agricultural Engineering Building,, Pennsylvania State University, University Park, PA 16801, (2) Agricultural and Biological Engineering, The Pennsylvania State University, 225 Agricultural Engineering Building, University Park, PA 16802-1909

Pretreatment of lignocellulosic biomass is both chemically and economically intensive. Improving the quality of biomass during storage can reduce the severity of pretreatment and benefit other downstream processes. Ensilage is anaerobic solid state fermentation, where the production of organic acids creates an acidic environment, typically pH 4.0. The severe conditions inhibit microbial growth and preserve biomass. Trichoderma reesei Rut C30 is a filamentous fungus that produces cellulases and xylanases when grown aerobically on lignocellulosic biomass at 30 C. The fungus converts polysaccharides to sugars but also consumes a considerable amount of biomass for its growth. Performing ensilage after obtaining some polysaccharide conversion may control both polysaccharide conversion and biomass loss. Two different experiments with 7 and 14 days of aerobic phase were conducted. The effect of sterilization of biomass prior to fungal growth was studied with 7 days of aerobic phase. The effect of a higher temperature (50 C) on enzyme activity and polysaccharide conversion was studied with 14 days of aerobic phase. In both the experiments, ensilage was performed before and after fungal growth separately to determine the optimal order of anaerobic and aerobic phases. The processes with aerobic phase after ensilage consumed the organic acids produced increasing the pH levels back to 6.0. Higher temperature caused higher dry matter (DM) loss of about 0.4 g/g initial DM compared to 0.15 g/g initial DM. Surprisingly, sterilized biomass exhibited higher loss of dry matter. The best process was considered to be 7 days of aerobic phase before ensilage on unsterile stover, which produced similar polysaccharide conversion to 14 days of aerobic phase but significantly lower dry matter loss.

573k Fractionation and Optimization of Switchgrass Pretreatment Via Cellulose-Solvent-Based Lignocellulose Fractionation (CSLF) at Lower Cellulase Loadings

Noppadon Sathitsuksanoh, Zhiguang Zhu, and Y-H. Percival Zhang. Biological Systems Engineering, Virginia Polytechnic Institute and State University, 200 Seitz Hall (0303), Blacksburg, VA 24061

Lignocellulosic biomass as a sustainable source of biofuels and value-added renewable materials is hindered by the high processing costs and relative low sugar yields. Switchgrass is of great interest as a potential bioenergy feedstock. A new cellulose-solvent-based lignocellulose fractionation (CSLF) technology has been developed for separating lignocellulose components based on their different solubility in different solvents under modest reaction conditions (Zhang et al. Biotechnol. Bioeng. 2007. 97: 214). This technology is regarded as a nearly generic pretreatment addressing different types of feedstock with the high enzymatic glucan digestibility within a short reaction time. The optimal pretreatment conditions for switchgrass were determined to be 85% H_3PO_4 at 50ºC for 45 minutes. A 3-fold reduction in total enzyme loadings from 15 FPU of cellulase and 30 CBU of ß-glucosidase per gram of glucan to 5 FPU of cellulase and 10 CBU of ß-glucosidase per gram of glucan still lead to ~90% glucan digestibility within 24 hour hydrolysis. Scanning electron microscopy (SEM) images clearly show that CSLF can completely destruct fibril structure of recalcitrant lignocellulosic biomass. Quantitative surface area data show that CSLF can break lignocellulosic structure efficiently, resulting in a very high surface area of cellulose fraction and a decrease in the degree of crystallinity. Therefore, the pretreated switchgrass by CSLF possesses high enzyme hydrolysis rates and high sugar yields.

573l Development of Enzyme Magnetic Nanoparticle Complexes for Cellulose Hydrolysis

Hee Joon Park, Chemical Engineering, University of Wyoming, 1000 E. University Ave., Laramie, WY 82070 and **Patrick A. Johnson**, Chemical & Petroleum Engineering, University of Wyoming, Dept. 3295, 1000 E. University Ave., Laramie, WY 82071.

Enzyme nanoparticles show promise in the ability to have their properties tailored through immobilization, which provide advantages such as enhanced stability and facile separation from reaction mixtures. We are exploring two types of enzyme immobilization: cross-linking to a magnetic nanoparticle core and binding to stimuli responsive polymers. Cross-linking methods, while reducing the overall enzymatic activity, retained activity through multiple cycles and up to forty days. Enzyme immobilized magnetic nanoparticles of different sizes will be tested for activity, long-term stability and recoverability. We will elucidate the dependence of the size of the complexes on activity and recoverability. Magnetic nanoparticles, varying in size from 20 – 80 nm have been synthesized and characterized via TEM imaging. We are immobilizing cellulases to demonstrate the application of the immobilization concept on cellulose hydrolysis. Currently, a single cellulase is being used but future experiments will involve more complex mixtures of endo – and exo-cellulases for optimal hydrolysis.

573m Effect of Mixing on Enzymatic Hydrolysis of Cellulosic Biomass

Yongming Zhu, Brandon Emme, and Mads Torry-Smith. Novozymes North America, Franklinton, NC 27525

Enzymatic hydrolysis of cellulosic biomass requires physical contact of cellulolytic enzymes with cellulose chains before they can be depolymerized. From kinetic point of view, a properly designed mixing can improve the efficiency of enzymes by evenly distributing the enzymes onto the substrate and by promptly removing the inhibiting sugar products from enzyme-substrate complexes. So far, only limited information is available in literature on the effect of mixing on enzymatic hydrolysis. This study sought to better understand this issue using our cellulase enzyme and model as well as real substrates. The influences of mixing on the hydrolysis of different types of substrates were compared. The hydrolysis was also carried out using different types of mixing devices and different mixing schemes. The results will be presented and discussed.

573n Sporl for Robust Enzymatic Hydrolysis of Woody Biomass

J.Y. Zhu, USDA Forest Service, Forest Products Laboratory, One Gifford Pinchot Dr, Madison, WI 53716

This study established a novel process using sulfite pretreatment to overcome recalcitrance of lignocellulose (SPORL) for robust and efficient bioconversion of softwoods. The process consists of sulfite treatment of wood chips followed by mechanical size reduction using disk refining. The softwood substrate obtained showed excellent digestability, i.e., over 90% cellulose conversion of substrate with enzyme loading of about 14.6 FPU cellulase plus 22.5 CBU â-glucosidase per gram of od substrate after 48 hours hydrolysis. The glucose yield from enzymatic hydrolysis of the spruce substrate per 100 g of untreated od spruce wood (glucan content 43%) was about 37 g (excluding the dissolved glucose during pretreatment). The pretreatment altered the wood chips which reduced electric energy consumption for size reduction to only about 19 Wh/kg od untreated wood, or about 19 g glucose/Wh electricity. Furthermore, SPORL produced substantially low amounts of fermentation inhibitors, hydroxymethyl furfural (HMF) and furfural, of about 5 and 1 mg/g of untreated od wood, respectively. In addition, similar to better results were achieved when SPORL was applied to Red Pine. By building on the mature sulfite pulping and disk refining technologies already practiced in the pulp and paper industry, SPORL has very few technological barriers and risks for commercialization.

573o Novel Cell Surface Display on Lactic Acid Bacteria and Its Application to Lactic Acid Production from Starchy Materials

Kenji Okano[1], Qiao Zhang[2], Tsutomu Tanaka[3], Hideki Fukuda[3], and Akihiko Kondo[2]. (1) Department of Molecular Science and Material Engineering, Graduate School of Science and Technology, Kobe University, Kobe, Japan, (2) Department of Chemical Science and Engineering, Graduate School of Engineering, Kobe University, Kobe, Japan, (3) Organization of Advanced Science and Technology, Kobe University, Kobe, Japan

Poly lactic acid (PLA) is the one of the important agro-based plastic that can be produced from inexpensive renewable biomass resources including starchy materials. For expanded use of PLA, cost reduction of the lactic acid as a raw material for PLA is very important.

As a novel process for efficient lactic acid production, we propose the use of cell surface display system. This system is based on utilizing naturally occurring cell surface proteins as scaffolds and displaying functional proteins on various cell surfaces. While most of lactic acid bacteria (LAB) can not assimilate starch, by displaying amylase on the cell surfaces of LAB, saccharification of starch and fermentation can be sequentially carried out. Furthermore, since displayed enzymes can be easily separated with the host cells, repeated utilization of enzymes and LAB cells can be achieved. These features will let lactic acid production process to cost-efficient.

In this study, we developed a novel cell surface display system on various kinds of LAB using tandem repeats of the cA peptidoglycan-binding domain that allow efficient display of heterologous proteins. As a result, we succeeded development of the whole cell biocatalyst that possesses high enzyme activity. Finally, using α-amylase displaying LAB, direct and efficient lactic acid fermentation from starch was accomplished.

573p Bioethanol Production from Barley Hull Using Saa (Soaking in Aqueous Ammonia) Pretreatment

Tae Hyun Kim, Department of Agricultural and Biosystems Engineering, Iowa State University, 3101 NSRIC, Ames, IA 50011-3310, Frank Taylor, Eastern Regional Research Center, USDA-ARS-ERRC, 600 East Mermaid Lane, Wyndmoor, PA 19038, and Kevin B. Hicks, Eastern Regional Research Center, Agricultural Research Service, 600 Mermaid Lane, Wyndmoor, PA 19038.

Barley hull, a lignocellulosic biomass, was pretreated using aqueous ammonia, to be converted into ethanol. Barley hull was soaked in 15 and 30 wt.% aqueous ammonia at 30, 60, and 75°C for between 12 h and 11 weeks. This pretreatment method has been known as "soaking in aqueous ammonia (SAA)". Among the tested conditions, the best pretreatment conditions observed were 75°C, 48 h, 15 wt.% aqueous ammonia and 1:12 of solid:liquid ratio resulting in saccharification yields of 83% for glucan and 63% for xylan with 15 FPU/g-glucan enzyme loading. Pretreatment using 15 wt.% ammonia for 24–72 h at 75°C removed 50–66% of the original lignin from the solids while it retained 65–76% of the xylan without any glucan loss.

Addition of xylanase along with cellulase resulted in synergetic effect on ethanol production in SSCF (simultaneous saccharification and co-fermentation) using SAA-treated barley hull and recombinant E. coli (KO11). With 3% w/v glucan loading and 4 mL of xylanase enzyme loadings, the SSCF of the SAA treated barley hull resulted 24.1 g/L ethanol concentration at 15 FPU cellulase/g-glucan loading, which corresponds to 89.4% of the maximum theoretical yield based on glucan and xylan.

SEM results indicated that SAA treatment increased surface area and the pore size. It is postulated that these physical changes enhance the enzymatic digestibility in the SAA treated barley hull.

573q A Preliminary Investigation into the Relative Amount of Enzyme Deactivation from Enzyme-Substrate Interactions and Non-Specific Deactivation

Zhuoliang Ye and R. Eric Berson. Chemical Engineering, University of Louisville, Louisville, KY 40292

In the enzymatic hydrolysis of biomass, the hydrolysis reaction undergoes a significant decrease in rate and does not reach 100% conversion even after several days of reaction time. One reason may be attributed to enzyme deactivation due to a variety of mechanisms such as thermal, mechanical, and/or chemical activity. Preliminary work was performed here to differentiate between loss of enzyme activity from enzyme-substrate interactions and other interactions of the enzyme in solution that are unrelated to the substrate (termed here as non-specific deactivation). To accomplish this, a series of tests were performed to compare cellulase's activity loss towards freshly added substrate following initial incubation of the enzyme in a buffer-only solution and in a buffer-substrate solution. The results have separated enzyme deactivation into two parts: deactivation due to enzyme-substrate interactions and non-specific deactivation. For high crystalline substrates, the contribution of each is about 50/50. For lower crystalline substrates, the contribution is also about 50/50. Since overall deactivation was less for the lower crystalline substrate, it is inferred that the majority of the deactivation of cellulase comes from exoglucanase.

573r The Heterogeneous Catalyst System for the Conversion of Free Fatty Acid In Used Vegetable Oils for the Conventional Production of Biodiesel

Young-Moo Park[1], Kwan-Young Lee[1], Joon-Yeob Lee[1], Jin-Suck Lee[2], and Deog-Keun Kim[3]. (1) Chemical and Biological Engineering, Korea University, 1,5 Ka, Anam-Dong, Seoungbuk-Ku, Seoul, 136-701, South Korea, (2) Biomass Research Center, Korea Institute of Energy Research, Yuseong, Daejeon, Korea, Dae-Jeon, 305-343, South Korea, (3) Korea Institute of Energy Research, Korea Institute of Energy Research, Yuseong, Daejeon, Korea, Dae-Jeon, 305-343, South Korea

The biodiesel produced by the transesterification of vegetable oils (VOs) has recently become more attractive due to its environmental benefits and the fact it is made from renewable resources. As the consumption of petroleum continues to increase, so does the associated air pollution and accelerated rate of global warming due to the increase of atmospheric CO2. If biodiesel was used as an alternative fuel, the production of CO2 would not increase, as the CO2 emitted from biodiesel would be recovered during the production of biomass.

Biodiesel is generally produced by the transesterification of triglyceride to methyl esters with methanol using homogeneous catalyst such as sodium or potassium hydroxide dissolved in methanol. An environmentally benign process for the production of biodiesel from VOs using heterogeneous catalyst (Na/NaOH/ã-Al2O3) has been developed recently by us. In order to reduce the production cost of the biodiesel, used vegetable oil is suggested to be used as feedstock. However, used VOs contain free fatty acids (FFA), which can react with the base catalyst if it used as a feedstock of the transesterification without any pretreatment. It brings a loss of catalyst and a production of soap as by-product. Therefore, FFA should be removed or converted to inert or reactant to overcome these problems. In this work, we tried to convert used VOs using special treatment to 100% FFA and convert the FFA using heterogeneous acid catalysts to fatty acid methyl esters (FAME), which are the components of biodiesel themselves. The activities of several heterogeneous catalysts were tested on the conversion of FFA and FAME. This reaction experimented to change the temperature, stirring speed, amount of catalyst and reaction time, even abundant 100wt% FFA was tested in optimal reaction conditions. The optimization of reaction conditions was also carried out. The characterization of heterogeneous catalyst was performed by ICP, TPD and XRD analysis.

573s The Enzymatic Conversion of Glycerol to Glycerol Carbonate

Kerri A. Cushing, Archana Trivedi, and Steven Peretti. Chemical and Biomolecular Engineering, North Carolina State University, Raleigh, NC 27695

The recent growth of the biodiesel industry has produced excess glycerol for which there is no market. This surplus glycerol offers an interesting opportunity to produce biomass-derived raw materials and chemical intermediates that have previously been manufactured from petroleum. If the surplus glycerol can be converted into a higher value product, the economics biodiesel production would be enhanced. This project aims to convert glycerol to glycerol carbonate, a versatile chemical that may be used as a solvent or additive by itself or as a platform chemical for other industries.

Lipases, which catalyze ester synthesis reactions in low water environments, were evaluated as biocatalysts for a reaction between glycerol and a dialkyl carbonate. Several lipases, dialkyl carbonate reactants, and solvents were considered. The conversion of glycerol to glycerol carbonate can be effectively catalyzed by lipases, particularly Candida antarctica lipase B. A solvent is not required for this reaction, but it can positively affect lipase activity. The rate of reaction as well as the product specificity differs for different dialkyl carbonate substrates.

When reacting with dimethyl carbonate in the presence of tert-butanol, glycerol conversion reached 72% with over 99% of the product being glycerol carbonate. The effects of dialkyl carbonate choice and solvent selection on lipase activity and product selectivity will be presented.

573t Two-Phase Flow Behaviors In Microtube Reactors during Biodiesel Production

Katsuki Kusakabe, Guoqing Guan, Marion Teshima, and Nozomi Sakurai. Department of Living Environmental Science, Fukuoka Women`s University, 1-1-1 Kasumigaoka, Higashi-ku, Fukuoka, 813-8529, Japan

Biodiesel is considered to be a cleaner-burning diesel fuel made from natural, renewable sources such as vegetable oils or animal fats. Various processes have been developed to produce biodiesel. Transesterification of vegetable oils and animal fats using methanol to produce biodiesel is a two-phase reaction in which methanolysis occurs only in the methanol phase. However, oil has a limited solubility in methanol so that vigorous mechanically mixing and/or heating are necessary to improve the mass transfer between the two phases. To resolve this problem, we examined the application of microreactor process to biodiesel production because of the properties of short diffusion distance and large specific surface area.

Transesterification of sunflower oil with methanol by using KOH catalyst to fatty acid methyl esters (FAMEs) was performed in microtube reactors. Influenced of microtube diameter, methanol/oil molar ratio and reaction temperature on the oil conversions were investigated. Flow behaviors in the transparent microtube (inner diameter = 0.8 mm, length = 1000 mm) under different operating conditions were observed and characterized by optical measurements. Methanol was colored to get clear images of the flow patterns in the microtube by inert red phloxine B. The relationship between the flow pattern and the oil conversion was discussed.

The segmented two-phase flow of the red methanol and the nearly-colorless oil phases was observed in all regions of the microtube in the absence of the reaction. The segmented flow was also observed at the entrance region with low conversion of sunflower oil, and then fine red droplets were dispersed and circulated in the oil segments with the progress of the reaction. In the case of a methanol/oil molar ratio of 23.9 at 60oC, quasi-homogeneous phase was formed at the place about 300 mm apart from the inlet because of high miscibility of the produced FAME and glycerol in methanol solution. As a result of the measurement of the oil conversion at each point of the microtube, sunflower oil was found to be completely converted to FAMEs in the quasi-homogeneous phase.

573u Stochastic Modeling of the Biodiesel Production Process

Sheraz Abbasi and Urmila Diwekar. Bioengineering, University of Illinois at Chicago, 851 South Morgan Street, Chicago, IL 60607

This investigation will focus on using stochastic modeling techniques to study the uncertainties in the biodiesel production process and analyze their impact on the plant efficiency and process economics.

There are inherent uncertainties in the biodiesel production process arising out of feedstock compositions, operating parameters and mechanical equipment design and can have significant impact on the product quality and process economics. The uncertainties can be quantified in the form of probabilistic distribution function. Stochastic modeling capability can be implemented in the ASPEN process simulator to take into consideration these uncertainties and the output could be evaluated to determine plant efficiency.

Diwekar and Rubin [1] have used stochastic modeling to study complex integrated gasification combined cycle (IGCC) power systems and have showed the impact of uncertainties on the overall plant efficiency.

Biodiesel production involves a trans-esterification reaction where long chain fatty acids are converted to mono alkyl esters and glycerol. Sodium hydroxide is used as a catalyst. The glycerol and esters are separated out in a settling vessel. The alcohol is separated from the glycerol and ester by either evaporation or flashing. The ester is then neutralized, washed with slightly acidic water to remove methanol and then dried and stored. The uncertainties are in the biodiesel feedstock: soybean oil, canola oil and animal grease. As most of the biodiesel is produced in a batch process, there are time dependent uncertainties as well as variables in the operating parameters. The results of this study will show the impact of the uncertainties on biodiesel plant output and provide a useful simulation model which is more accurate representative of real world conditions.

1.Diweker, U.M. and E.S.Rubin (1990), Stochastic Modeling of Chemical Processes, Computers Chemical Engineering, Vol. 15, No. 2, pp 105-114, 1991

573v Synthesis and Analysis of a Flexible Multi-Feedstock Biodiesel Production Process

Grace Nworie[1], René D. Elms, and Mahmoud El-Halwagi[2]. (1) Chemical Engineering, Texas A&M University, College Station, TX 77845, (2) Department of Chemical Engineering, Texas A&M University, College Station, TX 77843

Biodiesel can be produced from a variety of feedstock. Oils from vegetable and plant sources have been commonly used as feedstock but other oil sources are also viable alternatives. Soybean, palm, sunflower, rapeseed, cottonseed, safflower, and canola oils are popular feedstock for current Biodiesel production plants. Animal fats, waste oils and even oils from microalgae are other renewable feedstock options for the transesterification of triglycerides to fatty acid methyl esters. With the current advances in research, the changing prices, and the supply, and demand of these feedstocks, there is a need to design flexible plants that can process various inputs. The objective of this work is to develop a systematic procedure for the design and operation of flexible biodiesel plants that can accommodate a variety of feedstocks. In particular, the proposed project will employ a holistic approach and a combination of process simulation, synthesis, and integration techniques. The developed simulation, integration, and optimization models along with the integrated experiments will provide a new platform that can systematically provide a complete strategy for innovative pathways for producing biodiesel. These new techniques will lay the foundations for fundamental research in the emerging area of sustainable biofuels. A case study is solved to demonstrate the merits and applicability of the devised procedure.

573w Kinetics of Uncatalyzed Ethyl Esterification of Oleic Acid

Tanawan Pinnarat, Department of Chemical Engineering, University of Michigan, 3074 Dow Building, Ann Arbor, MI 48109-2136 and Phillip E. Savage, Chemical Engineering, University of Michigan, 3074 Dow Building, Ann Arbor, MI 48109-2136.

Biodiesel is an environmentally friendly fuel, which can be used in an unmodified diesel engine. The current commercial process for biodiesel production usually proceeds via the base catalyzed transesterification of triglycerides with methanol. There is interest in developing new processes that would produce biodiesel at a lower cost. Non-catalytic processes, typically at supercritical conditions, are among those being investigated. In one such process for catalyst-free biodiesel production, triglyceride is first hydrolyzed to free fatty acids. These are then esterified to make biodiesel in the second step. We have investigated this second step by studying the esterification of oleic acid with ethanol. The experiments were conducted in quartz tube batch reactors to ensure the absence of unintentional metal catalysis, which has occurred in other studies. The samples were analyzed by high pressure liquid chromatography (HPLC). We explored the effects of temperature, phase behavior, molar ratio of ethanol to oleic acid, and water content in the feed mixture and used these results to determine the kinetics of the reaction. Three different conditions were used; 250oC f = 0.05, 250oC f = 0.26, and 230oC f = 0.56. The variable f is the fraction of the reactor that we fill with reactant. Low values of f lead to mostly gas phase systems at the reaction temperature. Likewise, high values of f lead to liquid phase reactions. The results showed that 90% conversion was achieved for 250oC f = 0.05 at 60 minutes. Lower conversion (80%) for the same temperature at higher loading was obtained at the same reaction time. For the lower temperature, 230oC, only 70% conversion was observed even at a longer time of 80 minutes. These results indicate that conversion increases as temperature increases and the reaction does not require high pressure to be completed. It may be possible to do the non-catalytic esterification well below the critical pressure, which could lead to a lower cost process. The molar ratio of alcohol to oil is another important variable in the esterification reaction. Since the reaction is reversible, more alcohol will drive the reaction to produce more product. Five different ethanol to oleic acid molar ratios (1:1, 3:1, 5:1, 7:1, and 10:1) were compared for all three reaction conditions (250oC f = 0.05, 250oC f = 0.26, 230oC f = 0.56) at a reaction time of 40 minutes. At the stoichiometric ratio (1:1), low conversion was observed as expected. We obtained 50% conversion for 250oC f = 0.26 and 230oC f = 0.56 and a slightly higher conversion of 60% for 250oC f = 0.05. For 230oC, the conversion was highest at 3:1 and the conversion decreased as the molar ratio increased. At 250oC, the conversion increased by 20% when the molar ratio increased to 3:1 and 5:1, respectively, for f = 0.05 and f = 0.26, respectively. Adding more ethanol did not have a significant impact on the conversion. The water content in the reactant mixtue was also investigated, since water is a product of this reversible reaction. The experiment was conducted at five different water contents (1%, 3%, 5%, 10%, and 15% by volume). The conversion was compared at a reaction time of 30 minutes and molar ratio of ethanol to oleic acid of 10:1 for every case. The results show that the reaction is sensitive to the water content especially for 250oC f = 0.26. Adding only 1% water results in a decrease in conversion by 10% and more than 20% decrease resulted with 5% water content. The same trend was observed for 230oC f = 0.56 but the decrease in conversion started at 3% water content. At 250oC f = 0.05, the reaction is more tolerant to water content. The conversion did not change significantly up to 3% water. The conversion decreased by 20% at maximum water content studied here. This research work shows that the esterification of oleic acid can be done non-catlytically at conditions milder than supercritical. The optimal condition is 250oC, f = 0.047, a molar ratio of 3:1 and with less than 3% water in feed. In addition, to presenting experimental results, we will report the kinetics of esterification and the importance of metal catalysis for this reaction.

573x Viscosity Measuring Techniques for Biomass Solids Slurries Using Experimental and Computational Methods

Rajesh K. Dasari, BioEnergy International, Woburn, MA 01801 and **R. Eric Berson**, Chemical Engineering, University of Louisville, Louisville, KY 40292.

The determination of the rheological behavior of biomass slurries is important for designing reactors and processes for conversion to cellulosic ethanol, especially at large-scale. This will become even more important as cellulosic ethanol moves closer to commercialization. The settling of biomass particles during viscosity measurements has always been a serious problem causing erratic results. Maintaining homogeneity by uniformly suspending all the solid particles is necessary for obtaining accurate measurements of the viscosity of particulate slurries.

Settling of solids is observed for solids concentrations less than ~12% for pretreated corn stover (PCS) solids, above which the instantaneous separation of solids and liquid phases does not occur. A method is introduced here that incorporates both the determination of the uniform suspended speed (USS) in a vane/cup rheometer and the time needed for premixing to reach a steady-state viscosity prior to viscosity measurements. The USS for various solids concentrations of PCS slurries was determined both experimentally and with computational fluid dynamics (CFD). CFD simulations also provide visual verification of the distribution of solids in the viscometer cup. The density of wet PCS solids, which is difficult to determine experimentally, is determined to be 1100 kg/m3 based on the volume fraction distribution of solids at the USS.

573y A Simplified Approach to Aspen-Plus© Mass and Energy Balances for Preliminary Process Analysis and Economic Feasibility of Pyrolysis of Switchgrass to Bio-Oil

Frank Taylor, Charles Mullen, Andrew McAloon, and Akwasi A. Boateng. Eastern Regional Research Center, USDA-ARS-ERRC, 600 East Mermaid Lane, Wyndmoor, PA 19038

Process costing and analysis is best accomplished with rigorous chemical process simulation software such as Aspen-Plus© (Aspen Technology, Cambridge, MA). Processes that convert biomass feedstock to liquid fuels are unfortunately difficult to simulate due to the chemical complexity and lack of data for thermodynamic and other properties of the feedstock and products. To overcome this problem, the pyrolysis of switchgrass to bio-oil was modeled in Aspen-Plus© using a simplified switchgrass composition and simplified set of pyrolysis and gasification reactions that could be adjusted to agree with measured data. (1) Thus modeled, the process was scaled-up and evaluated for feasibility. The simulation is potentially applicable to thermochemical conversion strategies that optimize production of any of the three products, syngas, bio-oil and char from biomass feedstocks.

(1) A. A. Boateng, D. E. Daugaard, N. M. Goldberg and K. B. Hicks. Bench-scale fluidized-bed pyrolysis of switchgrass for bio-oil production. Ind. Eng. Chem. Res. 46:1891-1897 (2007).

573z Process Optimization for Maximum Ethanol Production Using Different Ethanologens from Afex Treated Rice Straw

Cheng Zhong[1], Chandraraj Krishnan[1], Ming-Woei Lau[1], **Venkatesh Balan**[1], Bruce Dale[1], and Ying-jin Yuan[2]. (1) Chemical Engineering and Material Science, Michigan state University, 2527 Engineering Building, East Lansing, MI 48824, (2) Department of Pharmaceutical Engineering, School of Chemical Engineering and Technology, Tianjin University, Tianjin, China

There is a growing need to find other alternatives to crude oil as the primary feedstock for the chemicals and fuels industry. Ethanol has many desirable features as a petroleum substitute and could help make a smoother transition from a petroleum-based to a bio-based chemical economy. Ethanol is produced in large quantities from natural resources including corn grain and sugarcane juice. However, there is a need to find an inexpensive and widely available lignocellulosic source of biomass (i.e. corn stover, rice straw, wheat straw) to avoid feedstock conflict with the prevalent food industry. Lignocellulosic recalcitrance is one of the primary impediments in the successful implementation of a cellulosic ethanol based biorefinery. AFEX is a leading pretreatment process to pretreat biomasses which reduce this intrinsic recalcitrance and is critical to help improve bioconversion.

Rice straw is an important lignocellulosic biomass with nearly 800 million dry tons produced annually worldwide. In addition to glucan and xylan it has about 10-15% silica which could be valuable byproducts if processed after enzymatic hydrolysis. Rice straw therefore has great potential as a lignocellulosic feedstock for making renewable fuels and chemicals. We will present the complete process optimization which includes pretreatment, enzymatic hydrolysis and fermentation to produce maximum ethanol using different ethanologens like yeast and bacteria.

573aa Optimal Allocation of Carbon Resources In An Integrated Biorefinery

Norman E. Sammons Jr.[1], Wei Yuan[1], Susilpa Bommareddy[1], Mario R. Eden[1], Burak Aksoy[2], and Harry T. Cullinan[2]. (1) Department of Chemical Engineering, Auburn University, 230 Ross Hall, Auburn University, AL 36849-5127, (2) Alabama Center for Paper and Bioresource Engineering, 242 Ross Hall, Auburn University, AL 36849

The integrated biorefinery, which uses renewable feedstocks from the forest-based industries, has the opportunity to provide a strong, self-dependent, sustainable alternative for the production of bulk and fine chemicals from polymers, fiber composites and pharmaceuticals to energy, liquid fuels and hydrogen. With such a wide range of processing steps and possible products, it is obvious that identification of the optimum process structure can not be done based on heuristics or rules of thumb. Changing market conditions may dictate a dynamic optimum for the allocation of resources and production capacity, resulting in a myriad of possible long-term product portfolios as well as a need for a net present value perspective which takes into account the time value of money. Economic market analysis, predictive financial modeling, and optimization under uncertainty are tools that can be used to determine the sensitivity of a decision-making framework to market fluctuations. Thus there is a need for systematic, reliable methods capable of incorporating different levels of process detail in the decision making framework. In this work a mathematical optimization based framework is being developed, which enables the inclusion of profitability measures and other techno-economic metrics along with process insights and performance characteristics obtained from experimental and modeling studies. By utilizing process integration methods, the processing steps can be optimized to ensure efficient use of energy and materials resources while assuring an acceptable, minimal level of environmental impact through the use of EPA's WAR algorithm. An inherent benefit of the proposed framework comes from the decoupling of the complex models from the selection step, which results in the ability to adapt to new developments within any of the processing steps and thus also incorporate novel innovative production processes in the decision-making framework. In this way, experimental and theoretical efforts can supplement each other in a synergistic manner, by providing direction and data for continued work. This contribution will illustrate the strategy for developing the decision making framework as well as highlighting the flexibility of the framework to utilize data from technological breakthroughs in the field of biorefining.

573ab Life Cycle Assessment of Bio-Oil Production by Pyrolysis of Wood

Heather S. Thomas and Mark Bricka. Chemical Engineering, Mississippi State University, Box 9595, Mississippi State, MS 39762

The Department of Chemical Engineering is evaluating a pyrolysis process for the conversion of wood into bio-oil or pyrolysis oil for use in wood treatment. Historically, wood has been treated with a mixture of copper, chromium, and arsenic. Due to its toxic characteristics, CCA has been banned from use in the U.S. As older wood is being taken out of service, new methods of preservation are being explored for the wood being put into service. The resulting bio-oil from this pyrolysis process has shown promise as a wood preservative.

Before full scale production can begin a Life Cycle Assessment (LCA) will be performed on the process. LCA is a cradle-to-grave analysis involving the feedstock and materials of construction as well as storage, transportation, and disposal issues. The results of the analysis will be used to determine the economic viability of the process, the energy "break even point," and the carbon footprint of the process.

573ac Cellulase Production from Treated Lignocellulosic Agricultural Residues Supplemented with Paper Sludge

Li Kang[1], Wei Wang[2], and Y.Y. Lee[2]. (1) Chemical Engineering, Auburn University, 212 Ross Hall, Auburn University, Auburn, AL 36849, (2) Chemical Engineering Department, Auburn University, 212 Ross Hall, Auburn University, Auburn, AL 36849

The production of highly efficient cellulase is becoming one of the major challenges in converting lignocellulosic biomass to ethanol. In this study, the different ratio mixture of paper sludge with sulfuric acid treated corn stover was used as feedstock for cellulase production by Trichoderma Reesei RutC-30. We speculated part of ash content of sludge and lignin of acid treated corn stover was removed in the mixing process to decrease the influence of harmful factors to cell growth.The cellulase activity produced from water washed mixture was 4.0 FPU/ml which was much higher than that of acid treated sludge and water washed treated corn stover. The cellulase activity produced from sodium hydroxide washed mixture was 6.0 FPU/ml which was also much higher than that of sodium hydroxide washed treated corn stover. The production scheme proposed in this study represents an economical and efficient method for cellulase production.

573ad Sustainable Fuel for the Brazilian Transportation Sector

Renan P. Oliveira, Navneet R. Singh, Fabio Ribeiro, W. Nicholas Delgass, and Rakesh Agrawal. School of Chemical Engineering and Energy Center at Discovery Park, Purdue University, 480 Stadium Mall Drive, West Lafayette, IN 47907

Ethanol bio-refineries in Brazil use bagasse from sugarcane as source of heat and electricity, which makes this process very energy efficient. However, this process can be considered very inefficient from a carbon perspective. Only about a third of the carbon content in sugarcane is converted into ethanol through fermentation, the rest of it is lost as CO2 during fermentation itself and when bagasse is burnt for heat and electricity for the process. Here, we propose a novel integration of the hybrid hydrogen-carbon (H2CAR) process with sugarcane-ethanol production in Brazil. This H2CAR process uses bagasse and CO2 produced during fermentation as sources of carbon for the production of liquid hydrocarbon fuels. The excess heat of this process can be used to supply the entire thermal energy requirement of an ethanol bio-refinery.

The integration of these two processes increases the overall carbon efficiency (~100%), and the energy efficiency of the process is increased from 29.2% to 55%. The most important result of this integration is the reduction in the amount of biomass required to support the entire Brazilian transportation sector by a factor of 3. Sugarcane ethanol alone would require 2.1% of the total land area of Brazil in order to completely support the transportation sector, while the H2CAR process integrated with ethanol bio-refineries would require 0.78% of the total land area to grow sugarcane. This is less than what is harvested today in Brazil, a significant result that could one day make Brazil the world's largest exporter of sustainable fuels.

Reference:

"Sustainable fuel for the transportation sector," Rakesh Agrawal, Navneet R Singh, Fabio H. Ribeiro & W. Nicholas Delgass. PNAS, 104 (12), 4828-4833 (2007). http://www.pnas.org/cgi/content/full/104/12/4828

573ae Biorefinery Product Allocation Using a Flexible Optimization Framework

Norman E. Sammons Jr.[1], Wei Yuan[1], Susilpa Bommareddy[1], Mario R. Eden[1], Burak Aksoy[2], and Harry T. Cullinan[2]. (1) Department of Chemical Engineering, Auburn University, 230 Ross Hall, Auburn University, AL 36849-5127, (2) Alabama Center for Paper and Bioresource Engineering, 242 Ross Hall, Auburn University, AL 36849

The forest based industries of the nation possess tremendous unrealized potential for the production of multiple, high-value added products from renewable biomass resources. These products range from bulk and fine chemicals through polymers, fiber composites and pharmaceuticals to energy, liquid fuels and hydrogen. This concept of an integrated biorefinery has the opportunity to provide a strong, self-dependent, sustainable alternative for the production of chemicals and fuels. A fundamental requirement to achieve this vision is the availability of concentrated, low-cost, easily available biomass resources as inputs into the biorefinery. Changing market conditions may dictate a dynamic optimum for the allocation of resources and production capacity, resulting in a myriad of possible long-term product portfolios and a need for a net present value perspective which takes into account the time value of money. Economic market analysis, predictive financial modeling, and optimization under uncertainty are tools that can be used to determine the sensitivity of a decision-making framework to market fluctuations. By constructing simulation models for various biorefining processes, determining variable and fixed cost components of these processes, and combining this cost data with information on expected return and market conditions, firms interested in biorefining will be able to determine the effects of changes in shareholder value as a result of deciding which products to pursue among the many biorefinery possibilities. This can be done while assuring an acceptable, minimal level of environmental impact through the use of EPA's WAR algorithm. An inherent benefit of the proposed framework comes from the decoupling of the complex models from the selection step, which results in the ability to adapt to new developments within any of the processing steps and thus also incorporate novel innovative production processes in the decision-making framework. In this way, experimental and theoretical efforts can supplement each other in a synergistic manner, by providing direction and data for continued work. This contribution will illustrate the strategy for developing the decision making framework as well as highlighting the flexibility of the framework to utilize data from technological breakthroughs in the field of biorefining.

BIOMASS TO SYNGAS CATALYSIS-RENEWABLE PROCESS FOR FISCHER-TROPSCH LIQUID FUELS

Savvas Vasileiadis* and Zoe Ziaka

Chemical & Materials Engineering, ZIVATECH, 535 Plum str.,

Youngstown, Ohio, 44503, USA

The purpose of this AIChE-100 technical communication is to present recent advances on multifunctional and multiphase reactors, and corresponding catalytic reaction-separation lines. These include permreactor and permeator configurations for low carbon renewable hydrocarbons and biomass feedstocks for high-yield conversion to hydrogen and carbon oxide mixtures, and subsequently to Fischer-Tropsch liquid fuels.

We analyze and elaborate on new process designs and operations for steam and CO_2 low carbon-hydrocarbon and biomass reforming, and the water gas shift. This analysis includes work in the design of new inline experiments and process operations. Also, the best parameter selection and optimization of such experiments for the final design and operation of the reformer. The proposed operations and catalysis advancements are of recent significance in the area of multifunctional reactors, including environmentally benign hydrocarbon and biomass conversion and abatement to valuable syngas (such as hydrogen and hydrogen-carbon oxides mixtures), [1,3,5].

Simultaneously, the reaction products, such as synthesis gas (mainly the H_2 and CO product mixture) are used for direct chemical syntheses such as high-yield liquid fuels, [6,7]. Moreover, alternative reaction products such as H_2 and CO_2 mixtures, coming out of the reforming reactor can be used as well for instantaneous liquid fuel synthesis with the selection of the proper catalysts [1,5].

Further objectives of these studies is the optimal development of multifunctional and multiphase processes and reactors with increased processing capacity and efficacy, conversion, separation, and additional utilization (e.g., through recycling streams and side feeds) of primary and secondary biomass-feedstocks and their related byproducts [1,5].

Permselective and adsorptive type processes with permeators and reactors can be used as example, in both the early purification stages of low carbon hydrocarbons and biomass derived gases from various contaminants (such as, NH_3, H_2S, HCl, HBr) as well as in the main conversion and upgrading sections of these feedstocks. Various improved materials, including ceramics, high Tg-polymers, types of carbons and composite materials can be used in the above selective operations for the final recovery of H_2, and H_2, carbon oxide mixtures from the reacted exit streams. For low conversion reformers a membrane permeator, made usually by polymer membranes, is suitable to be used after the reformer for gas separation purposes [1,5,8]. The rejected from the permeator gases, such as hydrocarbons, methane and/or carbon monoxide streams can be recycled into the inlet of the reformer or gas shift reactor for increasing the process yield and efficiency via the unreacted gas utilization and conversion.

Corresponding Author: vasil120@yahoo.com, **+30-2310-275473**

Secondary reaction products, as example these of H_2 and CO_2 can be also converted directly into liquid fuels. However, in an alternative route, after the CO_2 condensation and removal, pure hydrogen (H_2) is recovered and used in respective energy (e.g., fuel cells) and synthesis (e.g., hydrogenation) applications [9].

Furthermore, we seek to design and operate multifunctional and multiphase reactors and processes which eliminate the greenhouse gases emissions from the biomass source. Carbon dioxide (CO_2) is a main component of abundant hydrocarbon feedstoks including biomass, and byproduct of all fossil and synthetic hydrocarbon combustion processing, and as such needs to be recovered and concentrated or utilized and converted in the downstream [1,3-5,8].

There are described and proposed processes which seek for an in-situ carbon dioxide conversion and abatement of its negative greenhouse effects by utilizing alternative biomass reaction routes and special catalysts in the permreformer or reformer, as the direct reforming of methane with carbon dioxide in an all dry catalysis reforming process. The synthesis gas product, is separated by various techniques, namely membrane processes and can be used directly for chemical synthesis or as continuous fuel in fuel cells [1-3,5]. Chemical synthesis routes include liquid fuels production which is under continuous investigation.

With these reaction routes, biomass from various sources including forest and agricultural sources can be a valuable renewable feedstock for liquid fuels. Special care is given for catalysts selection in the reformer and Fischer-Tropsch reactor to avoid deactivation from carbon deposition from long hours in the stream. Special nickel classes of low quality nickel catalyst enriched with earth metals are used in the reformer, while appropriate catalysts are selected and experimented for the liquid fuels reactor, [1,8]. The economics of the process is deemed competitive.

Related modeling of such multifunctional reactors for the described applications is under way to describe their functionality, operational range, and working capacity in detail. The models used, simulate relevant acquired data and predict conditions for best multifunctional module and process operation by variation of several intrinsic parameters. Kinetic and transport parameters used in modeling can be obtained experimentally by separate kinetic and transport experiments. The described reactors and process configurations and their proposed applications seek to perform multiple reactive and catalytic unit operations within a single or integrated module which make them advanced in comparison with up to now proposed and utilized conventional reactors, [5]. Additionally, these operations are environmentally friendly as they produce no adverse chemicals or toxic gases.

In addition, objective of the above pilot plant and modeling studies is turnkey process development for advancing biomass conversion and selectivity, with increased hydrogen and syn-gas generation and flowing capacity, and increased overall biomass conversion efficiency. Further target elements are improved biomass processing, with economic and environmental consideration and efficient utilization for the proposed reformer and liquid fuel reactor designs. Also, the continuous utilization of primary and secondary biomass feedstocks and their derived byproducts to yield liquid fuels [1,3].

References:

1) S. Vasileiadis and Z. Ziaka, *Chemical Engineering Communications*, 76, 247-253, (1999).
2) J.R. Rostrup-Nielsen, *Journal. of Catalysis*, 85, 31, (1984).
3) S.Vasileiadis and Z. Ziaka, *Chemical Engineering Science,* 59, 4853-4859, (2004).
4) Z. Ziaka and S.Vasileiadis, "Polymer membrane reactors for enhanced hydrocarbon conversion and upgrading", *AIChE annual meeting, Proceedings series*, paper #275c Nov. (1997).
5) S. Vasileiadis and Z. Ziaka, *Separation &Purification Technology Journal,* 34, 213-225, (2004).
6) Q. Qin and D. Ramkrishna, *Industrial &Engineering Chem. Research,* 43,12, 2912-2921, (2004).
7) T.Y. Park and G.F. Froment, *Industrial & Engineering Chem. Research,* 40, 4172-4186, (2001).
8) Z. Ziaka and S. Vasileiadis, *Chemical Engineering Communications*, 156, 161-200, (1996).
9) C.S. Wen, G.V.Chilingarian, and T.F.Yen, "Properties and Structure of Bitumen, Bitumen, Asphalts, and Tar Sands", *Elsevier Publ. Co.,* New York, 7, 155 (1978).

584a Investigating Adsorbate Structures of Cyclic C_4 Oxygenates from Biomass on Catalytic Metal Surfaces Using Spectroscopic Techniques

Clay M. Horiuchi, Stephen T. Marshall, Andrea S. Loh, and J. Will Medlin. Department of Chemical and Biological Engineering, University of Colorado, ECCH 111 UCB 424, Boulder, CO 80309

Numerous studies have shown that the properties of metal catalysts can in principle be fine-tuned by controlling the composition of the metal surface with high precision. The ability to design catalysts capable of high selectivity towards the conversion of a single functional group in a multifunctional molecule is a major objective for heterogeneous catalysis research. This need for high selectivity toward a single functional group may be of growing importance in efforts to establish biorefining operations, where biomass-derived multifunctional carbohydrates are key "building block" intermediates and must be converted to commodity chemicals. This work focuses on results from high resolution electron energy loss spectroscopy (HREELS) and temperature programmed desorption (TPD) experiments combined with selective use of density functional theory (DFT) calculations on single-crystal surfaces under ultrahigh vacuum conditions to study structure-property relations for a series of C_4 cyclic oxygenates on Pd and Pt (111) surfaces. The objective of this work is to identify methods to tailor surfaces that are able to selectively react one functional group in the multifunctional molecule.

Two types of cyclic probe molecules have been studied in particular: 3-membered epoxide rings (in which ring-strain is high and the character of the oxygenate function is therefore more reactive) and 5-membered furanone rings (in which the ring is relatively stable). Both the epoxides and furanones contain an unsaturated C=C bond; for many applications (including biorefining applications), it is desired to selectively hydrogenate the olefin while keeping the oxygenate functionality intact. In this contribution, we explore the role of surface structure and composition in dictating the reaction pathways for multifunctional C_4 cyclic oxygenates on key transition metal and bimetallic surfaces. Possible catalyst design strategies for improving selectivity in reactions of these molecules are discussed.

584b Aqueous-Phase Hydrogenation of Acetic Acid on Monometallic Catalysts. A Combined Experimental and Theoretical Study

Ye Xu, Center for Nanophase Materials Sciences, and Chemical Sciences Division, Oak Ridge National Laboratory, P.O. Box 2008, MS-6493, Oak Ridge, TN 37831-6493, Hakan O. Olcay, Department of Chemical Engineering, University of Massachusetts, Amherst, Goessmann Laboratory, N. Pleasant Street, Amherst, MA 01003, and George W. Huber, Chemical Engineering Department, University of Massachusetts - Amherst, 686 N. Pleasant St, 112 Goessmann Laboratory, Amherst, MA 01003.

Depletion of fossil fuel combined with environmental and political concerns with fossil fuels have accelerated research on technologies to convert biomass into liquid fuels, i.e. biofuels. Aqueous-phase hydrogenation (APH) reactions are crucial for biofuel production, including ethanol production by hydrogenation of fermentation products (organic acids) [1], gasoline production by hydrogenation of pyrolysis oils [2], and alkane production by aqueous-phase dehydration/hydrogenation of carbohydrates [3]. APH reactions involve hydrogenating a range of functionalities, including acids, sugars, alkenes, and aldehydes. The objective of this project is to develop a fundamental understanding for aqueous-phase hydrogenations of organic acids by a combined theoretical and experimental approach.

We have measured the catalytic activity of aqueous-phase hydrogenation of acetic acid on supported monometallic Ni, Cu, Ru, Rh, Pd, Ir, and Pt catalysts in a fixed bed reactor. Ru/carbon is by far the most active catalyst, with an apparent activation energy of 44 kJ/mol. It is also the most selective toward ethanol at moderate temperatures, with the selectivity topping at 74% at 175 °C. Rh and Pt are the next most active, but their activity is an order of magnitude lower than that of Ru, with the rest of the metals being still less active.

For additional mechanistic insight, we have investigated the hydrogenation of acetic acid on the monometallic surfaces using density functional theory (DFT) calculations. By determining the energetics and kinetics of probable elementary steps, and taking experimental evidence into account, we tentatively identify the formation of acetyl (CH_3CO) as the main rate-determining step in the conversion of acetic acid, and the binding energy of acetyl as an important parameter of the overall activity. The effect of the aqueous phase on the reaction mechanism will be discussed. This study will provide valuable insight for designing a more versatile, alloy-based APH catalyst in the future.

[1] T. Eggeman, D. Verser, Recovery of organic acids from fermentation broths, Appl. Biochem. Biotechnol. 2005, 121-124, 605.

[2] G. W. Huber, S. Iborra, A. Corma, Synthesis of transportation fuels from biomass: chemistry, catalysts and engineering, Chem. Rev. 2006, 106, 4044.

[3] G. W. Huber, J. A. Dumesic, An overview of aqueous-phase catalytic processes for production of hydrogen and alkanes in a biorefinery, Catal. Today 2006, 111, 119.

584c Millisecond Autothermal Steam Reforming of Cellulose by Reactive Flash Volatilization

Joshua L. Colby, Paul J. Dauenhauer, and Lanny D. Schmidt. Department of Chemical Engineering and Materials Science, The University of Minnesota - Twin Cities Campus, 421 Washington Ave SE, Minneapolis, MN 55455

Three biomass-to-liquid processes (volatilization of cellulose, tar cleaning of organic products, and water-gas-shift of the gaseous effluent) have been combined into one autothermal catalytic reactor for the production of synthesis gas at millisecond residence times. Cellulose was reacted on the hot, catalytic bed of Rh-Ce/γ-Al2O3 catalyst on 1.3 mm α-Al2O3 spheres in the presence of O2, N2, and steam in a continuous flow fixed-bed reactor at 500-1100 °C. Complete fuel conversion to gases was observed for all experiments, in which N2/O2 ratio, S/C ratio, cellulose flow rate, and C/O ratio were varied. Steam addition increased the selectivity to H2 and decreased the selectivity to CO in agreement with water-gas-shift equilibrium. The process was demonstrated with ~80% selectivity to H2 at a synthesis gas ratio of H2/CO=2.3 with a pure oxygen gas feed at a fuel efficiency of ~75%. Carbon-free processing was explained by thermodynamic predictions for the formation of solid carbon, CS, at the observed experimental operating conditions.

584d Characterization of Ni-Based Catalysts Used for Steam Reforming of Simulated Biomass Gasification Vapors

Matthew M. Yung, Whitney Jablonski, and Kimberly A. Magrini-Bair. National Bioenergy Center, National Renewable Energy Laboratory, 1617 Cole Blvd., MS 3322, Golden, CO 80401

Biorefineries that synthesize fuels and chemicals from sustainable, renewable resources can employ syngas chemistry to create a slate of products. The gasification of cellulosic biomass to create this syngas is an active area of research. This syngas, however, may contain tars and other poisons that should be reformed and/or removed in order to protect downstream synthesis catalysts and to efficiently utilize the biomass carbon and hydrogen.

The current contribution uses simulated gasification vapors to examine the steam reforming activity, deactivation, and regeneration processes of catalysts in a bench scale, plug flow reactor. Bulk and surface catalyst characterization techniques are used to probe the principle catalyst features affecting reaction activity, selectivity, and stability differences amongst the samples, which mainly consist of Ni-based materials. These results will be discussed in the context of next generation catalyst design.

584e Techno-Economic Study of the Production of Mixed Alcohols from Lignocellulosic Biomass Using Direct Gasification

Abhijit Dutta, Biorefinery Analysis and Exploratory Research, National Renewable Energy Laboratory, 1617 Cole Blvd, MS 3322, Golden, CO 80401

Ethanol is currently viewed as a promising liquid fuel alternative to gasoline. The thermochemical route for the production of mixed alcohols from lignocellulosic biomass via direct gasification is presented. The process involves the use of high-pressure oxygen blown gasifier to produce syngas; reforming and cleanup of the syngas, followed by synthesis to mixed alcohols. This techno-economic evaluation studies the minimum ethanol selling price from such a process. It looks at the cost distribution across various sections of the process. Technological improvements are necessary in the areas of tar reforming and alcohol synthesis to bring down processing costs. Efficient and cost-effective catalysts for these operations will be the most important areas of development. Sensitivity of production cost to developments in these areas is discussed. Improvements are also expected in the design and reliability of gasifiers. Other significant parts of the process are already mature and are used in commercial processes for other products. With the maturity of technology and achievement of projected improvements, feedstock will be the single largest cost component of lignocellulosic ethanol. The Air Separation Unit will always be a cost burden on the oxygen-blown gasification process.

Partial oxidation of ethanol over supported Ni catalysts: Effect of support

Sitthiphong Pengpanich[1], Vissanu Meeyoo[2] and Thirasak Rirksomboon[3]

[1] Department of Chemical Engineering, Mahanakorn University of Technology, Bangkok 10530, Thailand
[2] Centre for Advanced Materials and Environmental Research, Mahanakorn University of Technology, Bangkok 10530, Thailand
[3] The Petroleum and Petrochemical College, Chulalongkorn University, Bangkok 10330, Thailand

Abstract

In this study, the effect of supports on the catalytic activity of Ni based catalysts for the partial oxidation of ethanol was investigated in a fixed bed reactor over the temperature range of 300-650°C. The results showed that the product distribution depended on the type of supports. At C/O molar ratio of 1, the main products of the partial oxidation reaction were H_2 and CO_2 for all catalysts. Ni/$Ce_{0.75}Zr_{0.25}O_2$ catalyst presents the highest catalytic activity and selectivity towards H_2. The amount of carbon deposited on Ni/$Ce_{0.75}Zr_{0.25}O_2$ catalyst was less than the other catalysts due to its good redox properties.

Introduction

Hydrogen is forecasted to become a major energy carrier for the future since it can reduce pollution and greenhouse gas emissions. Therefore, the large amounts of hydrogen will be required in a hydrogen economy.

Hydrogen can be produced from hydrocarbons such as natural gas, naphtha, vacuum residue, refinery off-gas, gasoline, diesel, methanol etc. Among these, ethanol is a very promising candidate because it is a renewable raw material that can be obtained from biomass and addresses the CO_2 emission. Furthermore, the infrastructure needed for ethanol production and distribution has already been established in Thailand for producing gasohol.

There are three main methods for producing hydrogen from HCs, namely steam reforming, partial oxidation and autothermal reforming. Partial oxidation of ethanol is an attractive process for hydrogen production because it enables fast start up and does not need the addition of heat. However, this process has some disadvantages such as formation of by-products and catalyst deactivation (Fatsikostas and Verykios, 2004). Hence, the catalysts which are highly active, selective and stable under reaction conditions are needed. Nickel-based catalysts are attractive for partial oxidation of HCs due to their low cost. Since support plays a significant role on performance of supported metallic catalysts on this reaction (Ni et al., 2007; Mattos and Noronha, 2005), the study concerning the effects of metal and support on partial oxidation of ethanol is needed.

In this study, Ni supported on different supports (Al_2O_3, $Ce_{0.75}Zr_{0.25}O_2$, ZrO_2 and MgO) was investigated for the partial oxidation of ethanol in a fixed bed reactor over the temperature range of 300-650°C.

Experimental

Catalyst Preparation

The $Ce_{0.75}Zr_{0.25}O_2$ and ZrO_2 used as supports were synthesized by urea hydrolysis method as reported in Pengpanich et al. (2002) while commercial α-Al_2O_3 and MgO supports were supplied from Alfa Aesar.

The catalysts were prepared by incipient wetness impregnation method using an aqueous solution of $Ni(NO_3)_2.6H_2O$ (99.0%, Fluka). The amount of Ni loading was controlled at 5 wt %. The catalysts were then calcined in air at 500°C for 4 hr.

Catalyst Characterization

The catalysts were characterized by H_2 pulse chemisorption, XRD, and H_2-TPR. The amount of carbon deposition on the spent catalyst was quantified by TPO technique using 2% O_2 in He.

Catalytic Activity Tests

Partial oxidation of ethanol was carried out in a packed-bed quartz (i.d. 6 mm) reactor. Typically, ca. 0.1 g of the catalyst was diluted with 0.1 g of α-Al_2O_3. The Ethanol was vaporized from a saturator at 40 °C using Ar as carrier gas and mixed with air to obtain the desired C/O molar ratio of 1.0. The total flow rate of feed gases was kept at 60 ml/min (GHSV = 36,000 ml/h-g of cat) using Aalborg mass flow controllers. Measurements were carried out at furnace temperatures adjusted sequentially to 400 - 650°C. The gaseous products were chromatographically analyzed to separate H_2, O_2, N_2, CO, CO_2, CH_4, C_2H_4, acetaldehyde and acetone.

Results and Discussion

The main products of ethanol partial oxidation are H_2, CO_2 and CO with a trace of CH_4, C_2H_4, acetaldehyde and acetone for all catalysts. The ethanol conversion and product distribution over these catalysts are shown in Figure 1. The results showed that $Ni/Ce_{0.75}Zr_{0.25}O_2$ catalyst presents the highest catalytic activity and selectivity towards H_2 than the other catalysts. This might be due to the highest metal dispersion of this catalyst.

The product selectivities are varied depending on the catalysts. At temperature below 450°C, acetaldehyde was found in most catalyst systems and decreased with increasing temperature. This suggested that the dehydrogenation takes place over these catalysts at low temperature. It was found that a significant amount of acetaldehyde was obtained over Ni/MgO and Ni/ZrO_2 catalysts suggesting that dehydrogenation reaction of ethanol is more favour on MgO and ZrO_2 supports. Above 450°C, the amount of acetaldehyde was relatively low and other products including CO, H_2 and CO_2 products were more pronouced. This suggests that the partial oxidation was the main reactions at temperatures above 450°C.

As shown in Table 1, the amount of carbon deposition varies with type of supports decreasing in the order of Ni/α-Al_2O_3 > Ni/ZrO_2 > Ni/MgO > $Ni/Ce_{0.75}Zr_{0.25}O_2$. This might be due to the fact that $Ce_{0.75}Zr_{0.25}O_2$ has the highest oxygen storage capacity and good reducibility.

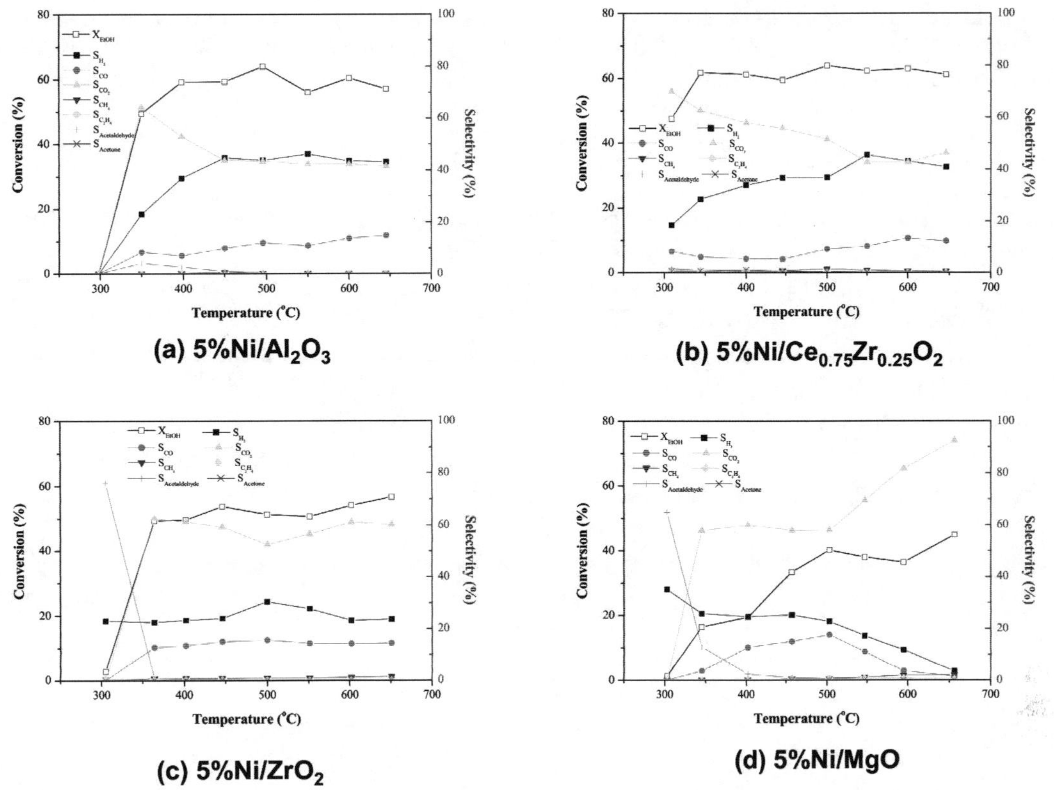

Figure 1 Ethanol conversion and product selectivities over 5 wt% Ni supported catalysts at the condition of GHSV = 36,000 ml/h-g of cat and feed molar ratio of C/O ratio = 1.

Table 1 The Amount of Carbon Deposition Quantified by TPO over The Spent Catalysts after 6 hr of Reaction at 500°C and C/O Ratio of 1.0

Catalyst	EtOH Conversion[1] (%)	Amount of carbon deposition (wt%)
Ni/$Ce_{0.75}Zr_{0.25}O_2$	59.5	15.2
Ni/ZrO_2	50.7	28.6
Ni/α-Al_2O_3	56.0	42.3
Ni/MgO	36.4	13.7

[1]at the end of reaction time (6 h)

Conclusions

It can be concluded that the catalytic activity of partial oxidation of ethanol depended on the nature of support. The main products of the reaction were H_2 and CO_2 for all catalysts. Ni/$Ce_{0.75}Zr_{0.25}O_2$ catalyst presents a good catalytic activity and selectivity for ethanol partial oxidation. The amount of carbon deposition on the Ni/$Ce_{0.75}Zr_{0.25}O_2$ catalyst was less than the other catalysts due to its good redox properties. This suggests that $Ce_{0.75}Zr_{0.25}O_2$ is a promising support for partial oxidation of ethanol.

Acknowledgements

The authors would like to thank National Metal and Materials Technology Center (MTEC) for the financial support.

References

Fatsikostas, A.N. and Verykios, X.E., "Reaction network of steam reforming of ethanol over Ni-based catalysts, " *J. Catal.*, **225**, 439 (2004).

Mattos, L.V. and Noronha, F.B., "Partial Oxidation of Ethanol on Supported Pt Catalysts," *J. Power Sources*, **145**, 10 (2005).

Ni, M., Leung, D.Y.C., and Leung, M.K.H., "A Review on Reforming Bio-ethanol for Hydrogen Production," Int. J. Hydrogen Energy **32**, 3238 (2007)

Pengpanich, S., Meeyoo, V., Rirksomboon, T., and Bunyakiat, K., "Catalytic Oxidation of Methane over CeO_2-ZrO_2 Mixed Oxide Solid Solution Catalysts Prepared via Urea Hydrolysis," *Appl. Catal. A*, **234**, 221 (2002).

617a Process Modeling Results of Various Leading Pretreatment Technologies for Corn Stover and Hybrid Poplar

Tim Eggeman, Neoterics International, 2319 S. Ellis Ct., Lakewood, CO 80228 and Richard Elander, National Bioenergy Center, National Renewable Energy Laboratory, 1617 Cole Boulevard, Golden, CO 80401.

The Biomass Refining Consortium for Applied Fundamentals and Innovation (CAFI) was formed in 1999 as a multi-institutional effort to develop comparative information on cellulosic biomass pretreatment options. Initial work focused on a comparative evaluation of corn stover feedstock for five leading pretreatment technologies (dilute acid, hot water, ammonia recycle percolation, ammonia fiber explosion and lime). More recently completed efforts, known as the CAFI 2 project, included both corn stover and hybrid poplar as feedstocks on which performance data has been collected. Also, an additional pretreatment approach (SO2 pretreatment), has been included in the research and process modeling aspects of the CAFI 2 project. Process modeling updates include the use of a more current costing basis for the economic models, the development of a graphical user interface, and the incorporation of a woody biomass feedstock handling system that is suitable for handling the hybrid poplar feedstock. The final pretreatment, enzymatic saccharification, and hydrolyzate fermentation data on corn stover and hybrid poplar for each pretreatment process has been incorporated into the economic models. Model scenarios that project a Minimum Ethanol Selling Price (MESP) for a "current" technology case (based upon available bench-scale data) and a "target" case (based upon reasonable projections of improved, commercial-scale performance) have been developed for each pretreatment process for both corn stover and hybrid poplar.

617b Short-Term Lime Pretreatment: Woody Vs Herbaceous Biomass

Rocio Sierra and Mark Holtzapple. Chemical Engineering, Texas A&M University, Chemical Engineering, 3122 TAMU, College Station, TX 77843

Lime pretreatment has demonstrated advantageous to importantly increase the digestibility of lignocellulosic biomass with little or no degradation of sugars. It has been amply studied at gentle conditions (up to 65°C and atmospheric pressure for 1 to 4 weeks) and more severe conditions (up to 180°C and 300 psig pressure with oxygen for 1 to 10 hours). However, no data has been recorded for short pretreatment at pretreatment times less than an hour. In this work, lime pretreatment is used on switchgrass and pine covering a temperature range between 100 and ~250°C using oxygen and pretreatment times between 5 min and 6 hours.

617c Pretreatment of Switch Grass by Aqueous Ammonia

Y. Y. Lee[1], V. Ramesh Pallapolu[2], and Urvi D. Kothari[2]. (1) Department of Chemical Engineering, Auburn University, 207 Ross Hall, Auburn University, AL 36849, (2) Chemical Engineering Department, Auburn University, 207 Ross Hall, Auburn University, AL 36849

Switch grass is one of the most promising energy crops for production of cellulosic ethanol. In this study, pretreatment of switch grass by aqueous ammonia was investigated applying two different modes of operation: ammonia recycle percolation (ARP), and soaking in aqueous ammonia (SAA). ARP is operated at high temperature (150°C-180°C) and short residence time (10-20 minutes) in a flow-through reactor, whereas SAA is carried out at low temperature (below 70oC) and long reaction time (6-24hrs) in batch reactor. The pretreated samples were analyzed for composition and subjected to enzymatic digestibility test using commercial cellulases with and without supplementation of xylanase. The results were analyzed to assess the overall effectiveness of the pretreatment and to determine the optimum conditions. Selected samples were further investigated for crystallinity, surface area, and surface composition by XRD, FTIR, and TGA in an attempt to verify the effects of these physical properties on the enzymatic reaction. Most of hemicellulose content is retained after pretreatment by aqueous ammonia due to high selectivity in ammonia reaction with lignin over carbohydrates. Simultaneous saccharification and co-fermentation (SSCF) was applied for optimally treated switch grass samples using recombinant E. coli (ATCC-55124) and aforementioned cellulase preparations. When it is properly operated, SSCF converts both glucan and xylan contents simultaneously in one stage. The simplicity of this process offers a significant economic benefit. The results of the SSCF test obtained from this work reaffirmed these claims.

617d Effect of Upland Vs. Lowland Variety on Sugar Yields Following Ammonia Fiber Expansion (AFEX) Pretreatment and Enzymatic Hydrolysis of Shawnee and Alamo Switchgrass

Rebecca J. Garlock, Venkatesh Balan, and Bruce E. Dale. Department of Chemical Engineering and Materials Science, Michigan State University, 2527 Engineering Building, East Lansing, MI 48824

Switchgrass (*Panicum virgatum* L.) is a native North American prairie grass and is considered by many to be the most promising dedicated energy feedstock for the production of lignocellulosic ethanol. Switchgrass cultivars can be divided into two distinct classes or cytotypes. Upland varieties are typically octoploid and found in higher, more mesic sites while lowland varieties are typically tetraploid and found at lower, more hydric sites. While there has already been a substantial amount of research into optimizing different pretreatment methods for switchgrass, including Ammonia Fiber Expansion (AFEX), research up until this point has not focused on potential differences that could arise between the two cytotypes. For this experiment, AFEX pretreatment was performed at a number of different conditions on two varieties of switchgrass, Shawnee (upland) and Alamo (lowland), followed by enzymatic hydrolysis to determine sugar yields. The composition of the enzyme cocktail used during enzymatic hydrolysis was also varied to obtain the optimal sugar yields.

617e Effect of Liquid Hot Water Pretreatment on Switchgrass Hydrolysis

Youngmi Kim, Laboratory of Renewable Resources Engineering (LORRE), Purdue University, Potter Engineering Center, 500 Central Drive, West Lafayette, IN 47907-2022, Nathan S. Mosier, Agricultural and Biological Engineering & Laboratory of Renewable Resources Engineering (LORRE), Purdue University, Potter Engineering Center, 500 Central Drive, West Lafayette, IN 47907-2022, and **Michael R. Ladisch**, Laboratory of Renewable Resources Engineering Department, Purdue University, Potter Engineering Center, 500 Central Drive, West Lafayette, IN 47907-2022.

The conversion of switchgrass to fermentable sugars and ethanol provides a cellulosic feedstock for production of fuel ethanol which may be grown on lands not suitable for food agriculture. Switchgrass itself consists of 33% cellulose, 25% hemicelluloses, 18% lignin, and 24% other. If switchgrass is processed without pretreatment, the maximal conversion achieved at an enzyme loading of 15 FPU/g glucan (5 FPU/g biomass) is less than 5%. When the switchgrass is pretreated in liquid hot water, the conversion increases by 25-fold, resulting in 80% glucose yield. The utilization of liquid hot water followed by enzyme hydrolysis and fermentation is described in this paper. The levels of enzyme loading and inhibition effects are briefly discussed as part of the overall CAFI research project.

617f Sugar Yields from Switchgrass for the Coupled Operations of Dilute Acid and Sulfur Dioxide Pretreatment and Enzymatic Hydrolysis

Jian Shi, Mirvat Ebrik, Bin Yang, and Charles Wyman. College of Engineering-Center for Environmental Research and Technology, University of California, Riverside, 1084 Columbia Ave, Riverside, CA 92507

Switchgrass has been highlighted as a leading herbaceous energy crop in the United States. As a part of a Biomass Refining Consortium for Applied Fundamentals and Innovation (CAFI) Project among leaders in biomass pretreatment, our research is focused on optimizing total sugar yields from dilute sulfuric acid and sulfur dioxide pretreatment followed by enzymatic hydrolysis of the pretreated solids. Dilute sulfuric acid and SO2 were applied to switchgrass at concentrations from 0 to 5% over a temperature range of 140-220°C for times of 0 to 60 minutes, and sugar release from hemicellulose and cellulose were measured for pretreatment and the subsequent enzymatic hydrolysis of the resulting solids. Combinations of cellulases and hemicellulases were employed to identify combinations of pretreatment conditions and enzyme formulations that give the maximum total sugar recovery with the lowest possible enzyme loadings/cost. These results also provide a better understanding of switchgrass bioprocessing and new insights that will facilitate lower cost processes and commercialization of cellulosic ethanol technologies.

636a High Yield Epoxidation of Fatty Acid Methyl Esters with Performic Acid Generated *In-Situ*

Alejandrina Campanella, Department of Chemical Engineering and Center for Composite Materials, University of Delaware, 150 Academy St, Newark, DE 19716 and Miguel A. Baltanás, Instituto de Desarrollo Tecnológico para la Industria Química (UNL - CONICET), Güemes 3450, Santa Fe, S3000GNL, Argentina.

In recent years, a worldwide interest in renewable resources from the biomass, has spurred the manufacture of biodiesel as a more benign fuel, capable of lower environmental impact. It is customarily produced by transesterification of the vegetable oil triglycerides usually with methanol, employing sodium methoxide (or hydroxide) as homogeneous catalyst. The resulting product is referred to as FAME (fatty acid methyl ester). Thus, FAMEs have become widely available commodities which, in view of the stringent requirements of the automobile industry, are produced with great purity. This, in turn, has reopened new avenues to oleochemicals synthesis. One of these alternatives involves transforming the double bonds of the fatty acid molecules into oxirane (epoxide) groups. Epoxides have a high commercial importance and are widely utilized in plastics manufacture, lubricants, detergents and –more recently- as intermediates in chemical reactions.

In this study soybean FAME was employed to analyze in detail the epoxidation reaction network when performic acid (PFA) is used as oxygen carrier. The epoxidation reaction and the parallel, deleterious reactions linked to the oxirane ring-opening. The influence of several process variables (viz., degree of mixing, temperature, mol ratio of FAME double bonds/hydrogen peroxide/formic acid and solvent addition) on the system was studied, seeking reaction conditions capable of achieving high epoxide yields. Concentrated H_2O_2 (60 wt%) was used throughout the study. Also, the relevant kinetic rate constants pertaining to this reacting system were found.

The reaction temperature has a high impact in the yield and selectivity of the system, but employing 40 °C it is possible to achieve high conversions and high yield as well. The epoxide yield is a strong function of the available amount of FA in the system. Formic acid is indeed crucial for producing enough PFA in the aqueous phase –which then passes to the organic phase to deliver an oxygen atom to the double bonds- but, also, is the main responsible for the opening of the oxirane rings. Also the H_2O_2-to-double bonds molar ratio was found to decrease the oxirane selectivity by ring-opening as said molar ratio was increased.

636c Product Distribution for Heterogeneous Catalytic Cracking of Acylglycerides on Commercial Catalysts

Tracy J. Benson[1], Rafael Hernandez[1], William T French[1], Earl E Alley[2], and William E. Holmes[2]. (1) Chemical Engineering, Mississippi State University, Mississippi State, MS 39759, (2) Mississippi State Chemical Laboratory, Mississippi State University, Mississippi State, MS 39759

Renewable fuels will be essential to sustain future energy demand, as petroleum supplies decrease and economies of developing nations continue to grow at a fast pace. Green diesel and green gasoline could be produced via cracking of crude lipids using catalytic processes currently employed in the petroleum refining industry. In contrast to biodiesel, green fuels could be distributed via the petroleum pipelines. Also, its production does not generate glycerine as a by-product.

In this work, model unsaturated lipids were reacted over Zeolite Y and silica-alumina solid catalysts at 400°C in a micro-bed reactor coupled with online GC/MS/TCD analysis. These model lipids were acylglycerols in which the fatty acid constituent was oleic acid, an unsaturated fatty acid commonly found in the fatty acid profile of many plant, animal, and microbial sources. Zeolite Y and silica-alumina are two commonly used catalysts within the petroleum industry.

The reactor used in this work was an in-house built cryogenic capillary catalytic cracker, called the Quatra C, which uses online mass spectrometry and thermal conductivity detectors. The online analysis via cryogenic oven cooling allowed for quick assessment of reaction products from C1 – C30 hydrocarbons, as well as, CO and CO2.

Initial studies on H-ZSM-5, a highly acidic benchmark catalyst, indicated that cracking of the unsaturated acylglycerides (mono-, di-, and triolein) begins at the double bond within the carbon backbone of the fatty acid constituent and proceeds with cyclization to produce a multiplicity of aromatic hydrocarbons and several low molecular weight hydrocarbon gases (C2 – C6). Cracking results using Zeolite Y showed a shift in the product yield and produced more low molecular weight unsaturated compounds and fewer aromatics. Silica-alumina, an amorphous catalyst, produced chiefly dienes (C5 – C9) and very few aromatic compounds. Results from this work suggest that triglycerides crack at much lower temperatures and light olefin gas yields are higher than for crude petroleum cracking. Therefore, process conditions must be altered from what are currently employed in petroleum refineries.

636d Catalytic Oxidation of Glycerol to High-Value Chemicals

Wenbin Hu and Arvind Varma. School of Chemical Engineering, Purdue University, 480 Stadium Mall Drive, West Lafayette, IN 47907

With increasing biodiesel production, large amounts of glycerol are obtained as byproduct, resulting in its price decrease from 43 to 10 cents/gallon in the last three years. Thus, it is important to find new ways to utilize glycerol, which we have used to produce dihydroxyacetone (DHA), valued at $20/lb, by catalytic oxidation. Various catalysts were tested to identify the one with the highest effectiveness. The catalytic kinetics are currently being obtained in both batch and flow reactors. The intrinsic kinetics will subsequently be utilized to develop a reactor model. After validation by comparing the calculated and experimental results, the model will be used to determine optimum reaction conditions for the maximum DHA yield, which in turn will be verified by experiments.

OXIDATION OF GLYCEROL ON DIAMOND COATED ELECTRODES

Peter Stehring, TU Graz, Graz, Austria
Susanne Lux, TU Graz, Graz, Austria
Thomas Hilber, BDI – BioDiesel International AG, Grambach, Austria
Peter Letonja, TU Graz, Graz, Austria
Matthäus Siebenhofer, TU Graz, Graz, Austria

Introduction

On the background of rising prices for fossil fuel and the implementation of the EU Directive 2003/30/EC "on the promotion of the use of biofuels or other renewable fuels for transport" the importance of alternative fuels like biodiesel or bioethanol increases.

Increasing production of biodiesel needs exploitation of byproducts. In case of biodiesel process the most important byproduct is glycerol[1]. Its price is subject of significant fluctuations because of a surplus of non food quality product. Several oxidation routes, especially the electrochemical oxidation, which enables formation of value added oxidation products for the chemical industry, seem to be a promising utilization.

Electrochemical oxidation is known as an environment-friendly ("green") process because chemical oxidation agents are substituted by electrons.

Target

Glycerol oxidation, subject of this ongoing project, has been widely investigated in the past. Several electrode materials, electrolyte compositions, pH-regions and temperature fields have been investigated by different research groups. To briefly sum up the results, the expected optimum set up for highly selective synthesis of a specific product has still not been found. Most electrode materials which have been tested in the past are not stable (e.g. silver [1], even gold anodes have limited stability [1]) or generate only cleavage products (e. g. nickel [1]). Only the use of a platinum anode in combination with a huge excess of sodium hydroxide as electrolyte provided acceptable yields of sodium lactate as product [2]. At our department [3] several electrode materials (e. g. DSA, glassy carbon) have been tested. Only diamond coated electrodes seem to be a promising choice for obtaining "better" results.

In the past mainly acids (like sulfuric acid) and alkali hydroxides were used as electrolytes. The use of acids is the easiest way, because acids are not consumed during the electrolysis and they provide good electric conductivity. So electrolysis works with low concentration of acid. Unfortunately at strong acidic conditions the cleavage of glycerol is the main anodic reaction and current yields are low. Alkali hydroxides provides better current yields but large excess is needed because alkali hydroxides are consumed by many reactions (aldol reaction, Cannizzaro reaction, CO_2-fixing ...) which lead to a complex product mixture. This complex product mixture and the huge amount of surplus sodium hydroxide make downstream processing a very difficult and expensive procedure. In conclusion metal salt electrolytes have been tested for their capability.

---

1 In the biodiesel production there is a yield of about 100 g glycerol per kg biodiesel.

Method and experimental setup

The electrochemical oxidation was carried out in batch operation in a thin layer pump cell (Figure 1). Anode and cathode compartment were separated by an anion exchange membrane. Anolyte and catholyte were pumped through the cell by peristaltic pumps (with a volume flow of 4 ml/s) in order to increase the mass transfer and to take the hydrogen out of the cell. The temperature was measured with a Pt100 resistance temperature detector and controlled with heat exchangers between pump and cell.

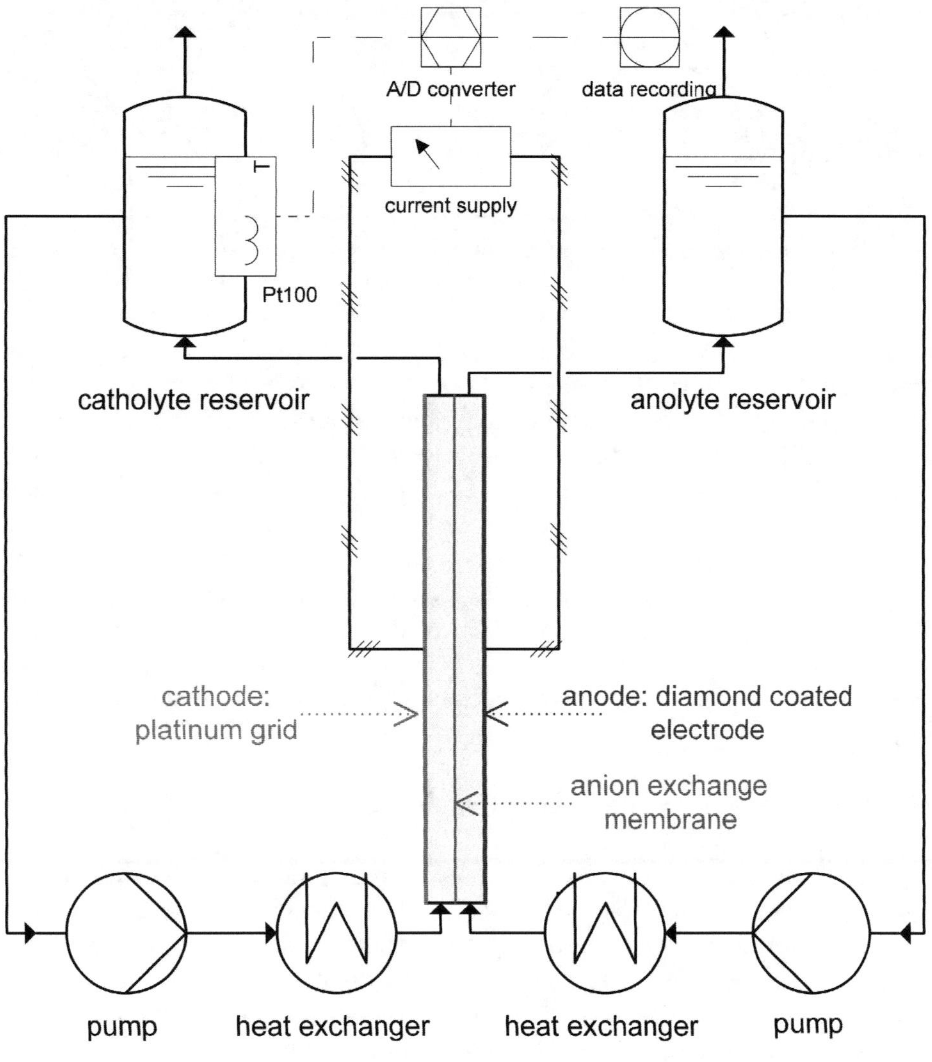

Figure 1. Sketch of the electrolysis cell

A diamond coated electrode (Fraunhofer IST) was used as anode. This electrode consists of a boron doped polycrystalline diamond layer on a niobium carrier.

Diamond coated electrodes have high chemical, mechanical, and thermal stability. The main advantage of diamond coated electrodes for anodic synthesis reactions is the distinct oxygen overvoltage. A platinum grid was used as cathode because of its very low overvoltage for hydrogen evolution, the only cathode product. All experiments were carried out with aqueous anolyte mixtures of glycerol ($c_{glycerol} = 0.25–2$ mol/dm^3).

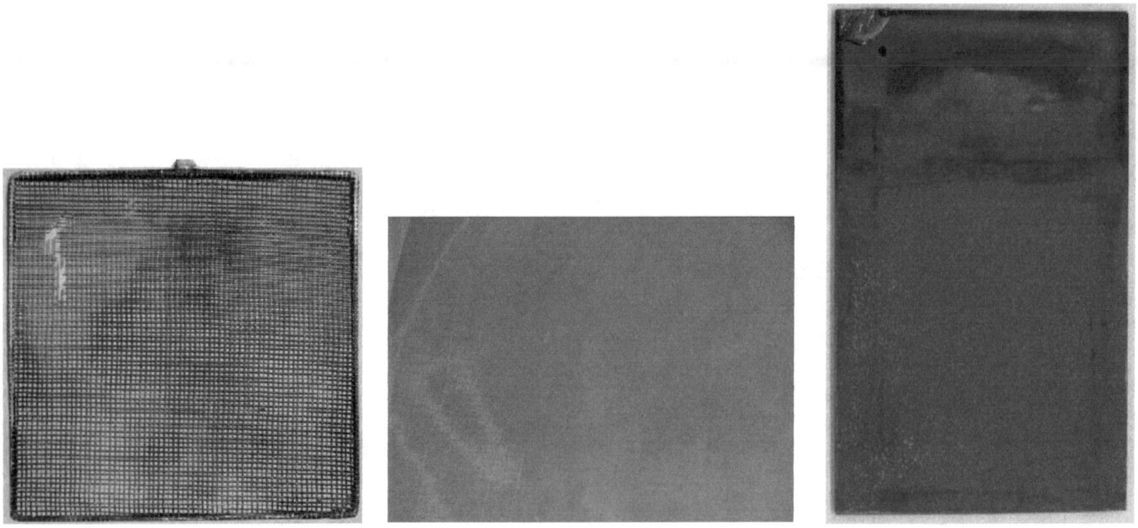

a b c

Figure 2. Images of the platinum anode (a), the anion exchange membrane (b) and the diamond coated anode (c)

The electrocatalytic activity of various electrolytes which provide different pH regions such as sodium sulfate (Na_2SO_4, moderate acidic because of organic acids which were formed during the glycerol electrolysis), *di*-potassium hydrogen phosphate (K_2HPO_4, neutral, buffer [$pK_A(H_2PO_4^-)$ = 7.207]) and sodium carbonate (Na_2CO_3, alkaline) were investigated. Aqueous solutions of sodium hydroxide (c_{NaOH} = 0.1 mol/dm^3) were used as catholyte. Thus charge transfer through the membrane was based on hydroxide ion transfer. During process optimization the current density (j = 75–305 A/m^2), temperature (θ = 20–50 °C), glycerol concentration and electrolyte concentration were varied. All experiments were carried out with constant current, flow and temperature. Cell voltage, current and temperature were recorded.

After specified time intervals samples were taken, acidified with dilute sulfuric acid and analyzed by HPLC.

Results

At the beginning of the electrolysis the desired products (glyceraldehyde and dihydroxyacetone) are formed with good selectivity and current yield. With ongoing process the fraction of cleavage products (glycolaldehyde and formic acid/formate) increases (**Figure 4**, **Figure 5**, and **Figure 6**). On a limited scale—especially at alkaline conditions—the oxidation of glyceraldehyde to glyceric acid and the oxidation of glycolaldehyde to glycolic acid take place (**Figure 3**). Through formation of acids during the electrolysis process the pH value decreases. The ratio between glyceraldehyde and dihydroxyacetone depends strongly on the pH value: At acidic conditions (Na_2SO_4 as electrolyte) only glyceraldehyde is formed, at neutral (K_2HPO_4 as electrolyte) and moderate alkaline (Na_2CO_3 as electrolyte) conditions mixtures of glyceraldehyde and dihydroxyacetone were formed.

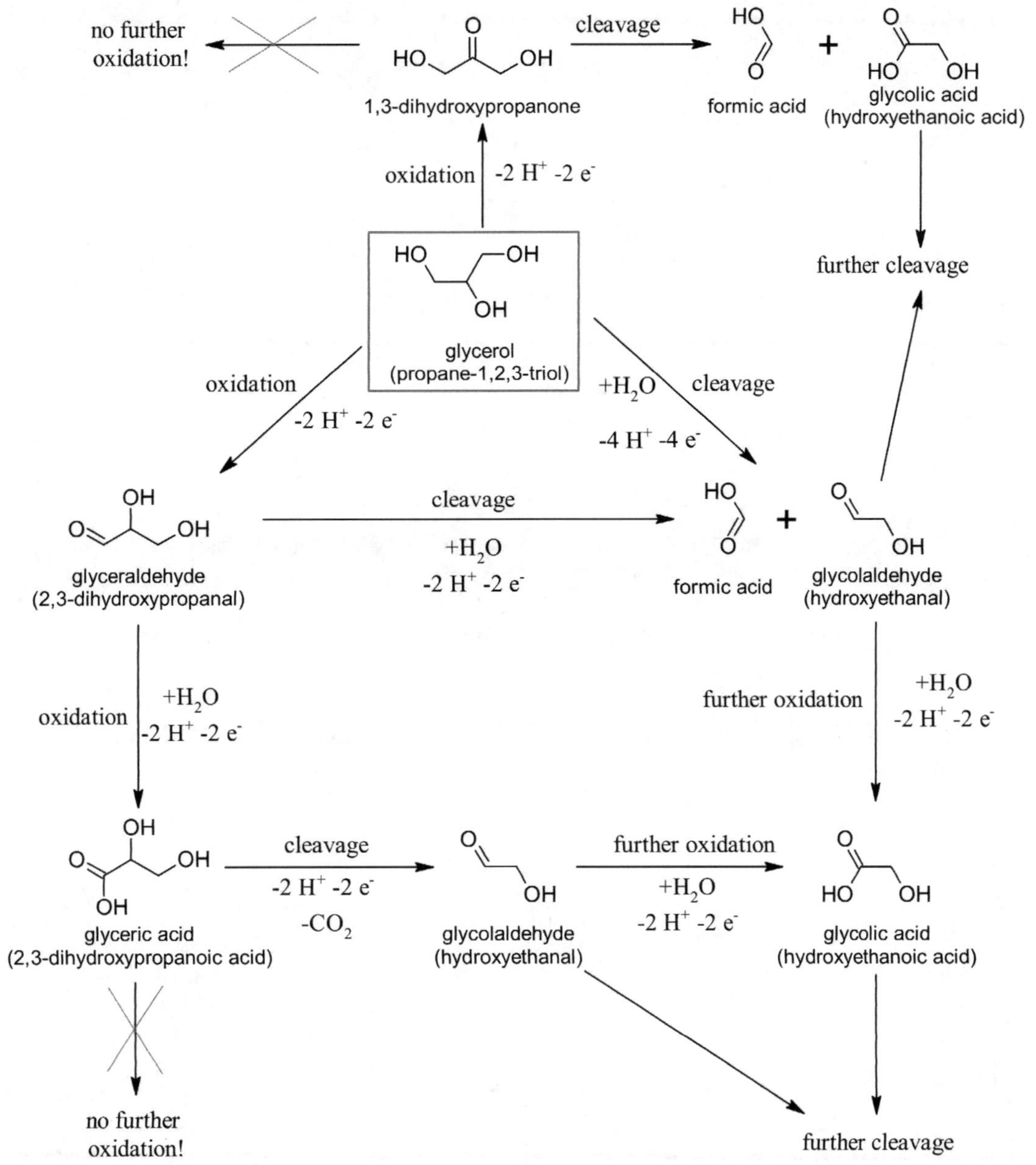

Figure 3. Overview of the anodic reactions of glycerol and its oxidation products without consideration of deprotonation reactions of the carboxylic acids

As a result of the cleavage reaction the concentration of glyceraldehyde and dihydroxyacetone reach a maximum. Through the decrease of the pH value dihydroxyacetone reaches its maximum prior to glyceraldehyde (cf. **Figure 5** and **Figure 6**).

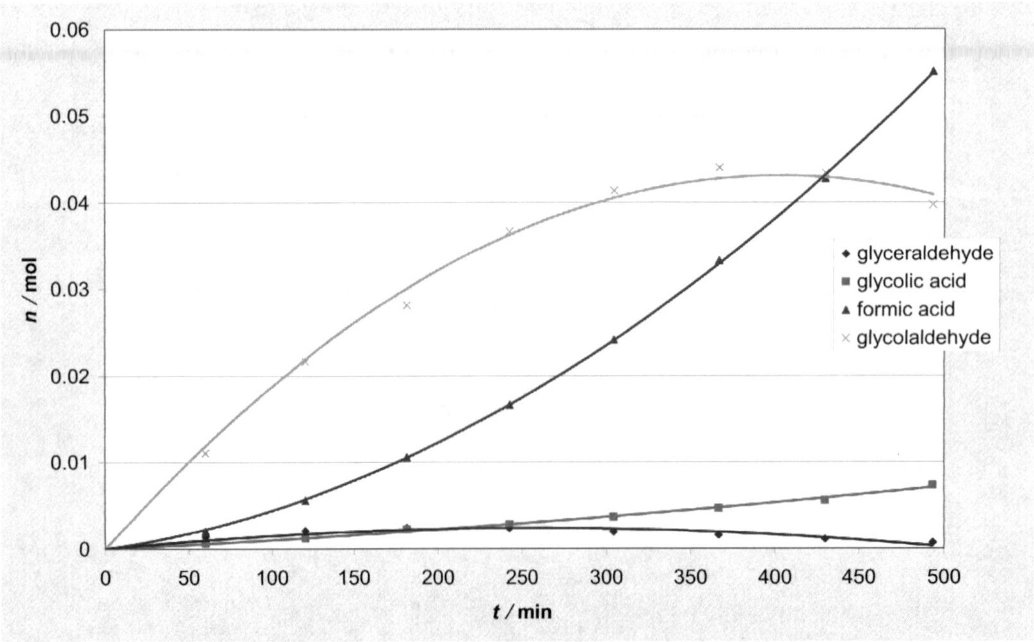

Figure 4. Formation of oxidation products versus time; electrolyte: Na_2SO_4, $\theta = 20$ °C, $j = 300$ A/m²

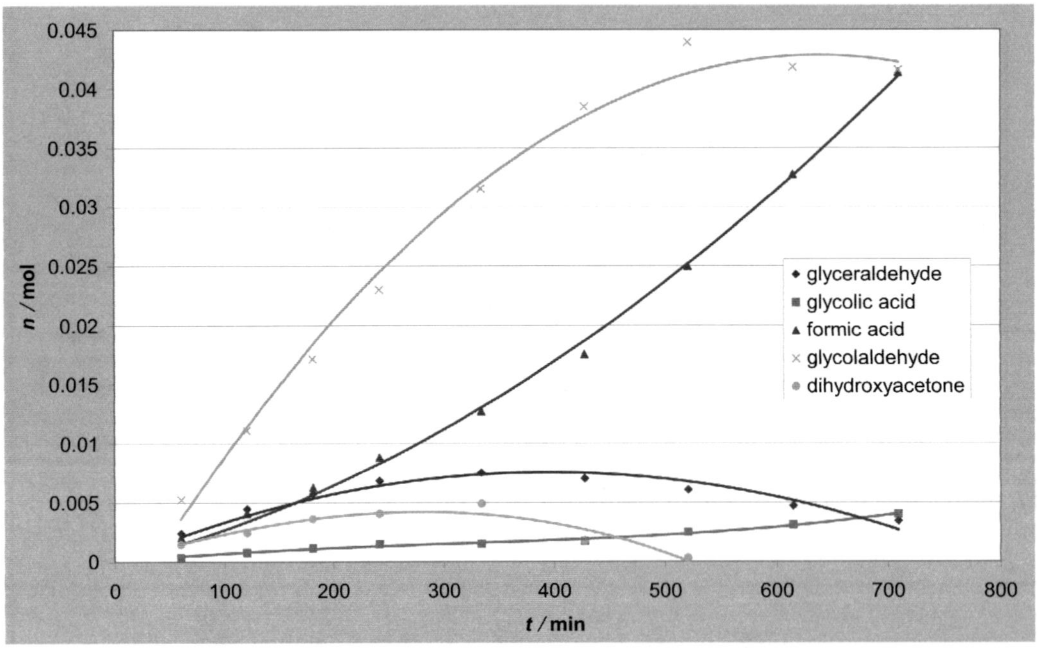

Figure 5. Formation of oxidation products versus time; electrolyte: K_2HPO_4, $\theta = 20$ °C, $j = 200$ A/m²

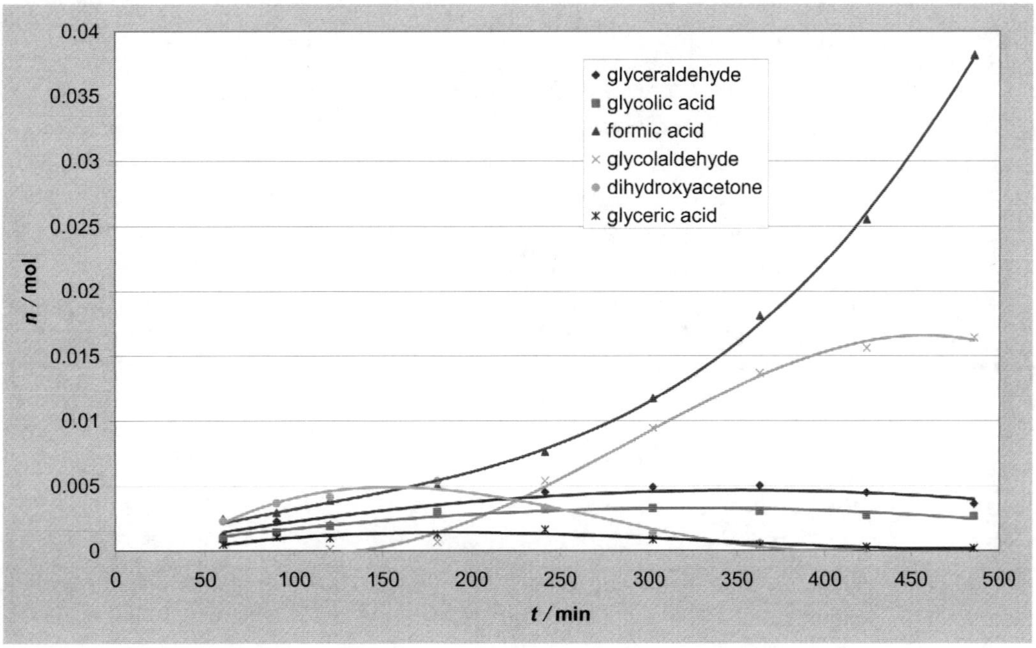

Figure 6. Formation of oxidation products versus time;
electrolyte: Na_2CO_3, $\theta = 20$ °C, $j = 305$ A/m²

The reaction kinetics of glycerol oxidation was determined (**Figure 7**). At the beginning of the process the degradation of glycerol follows zero-order rate law (electron transfer controlled reaction), with increasing electrolysis time a change to first order rate of degradation can be observed (mass transport controlled reaction). Formation of gaseous decomposition products (oxygen or carbon dioxide) is negligible.

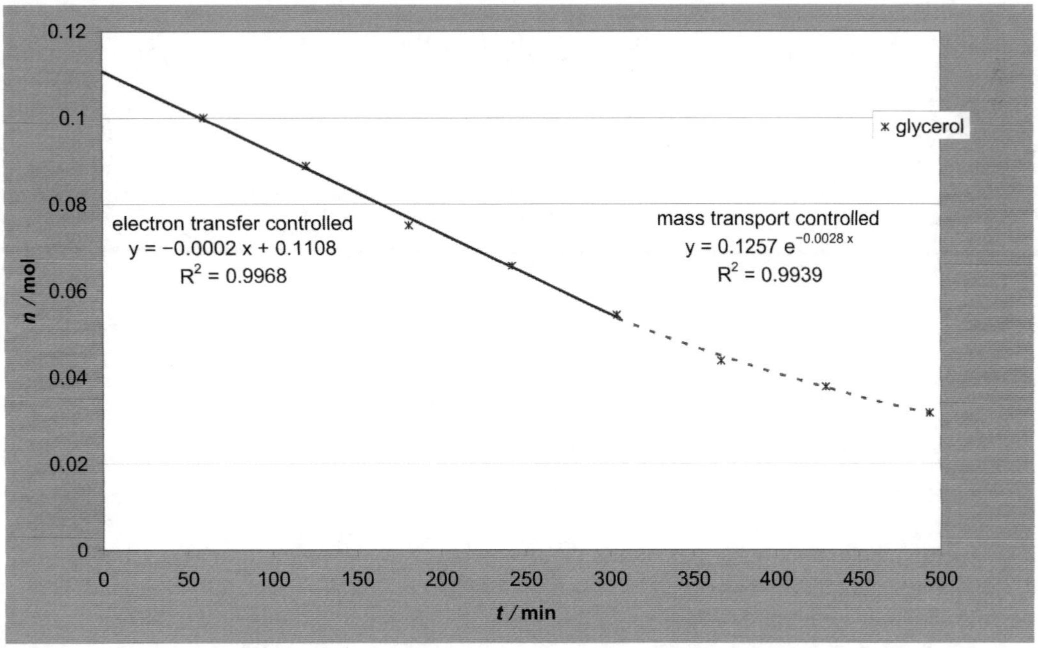

Figure 7. Glycerol versus time plot indicating electron transfer controlled reaction at the beginning and mass transport controlled reaction with ongoing time.
Electrolyte: Na_2SO_4, $\theta = 20$ °C, $j = 300$ A/m²

References

1. Weber, Ulrich (1993), *"Die elektrochemische Oxidation von Glycerin an unterschiedlichen Elektrodenmaterialien,"* Ph.D. Dissertation, University of Dortmund, Germany.
2. Tkaczuk, Piotr (1993), *"Elektrochemische Oxidation von Glycerin und Glycerinderivaten an einer Platin-Anode in verschiedenen Zelltypen,"* Ph.D. Dissertation, University of Dortmund, Germany.
3. Ernst, Martin (2007), *"Grundlagen und Optimierung elektrochemischer Synthesereaktionen am Beispiel der Oxidation von Glycerin,"* Ph.D. Dissertation, Graz University of Technology, Austria,

636f Conversion of Loblolly Pine Wood In Ionic Liquids

Carsten Sievers, Mariefel Valenzuela-Olarte, Teresita Marzialetti, Ildar Musin, Pradeep K. Agrawal, and Christopher W. Jones. School of Chemical & Biomolecular Engineering, Georgia Institute of Technology, 311 Ferst Drive NW, Atlanta, GA 30332-0100

Conversion of cellulose and wood in the ionic liquid 1-butyl-3-methyl-imidadzolium chloride has received considerable attention in the last years. The chloride anions of the ionic liquid disrupt hydrogen bonds between different polysaccharide chains resulting in the unique capability of this medium to dissolve cellulose. This property makes 1-butyl-3-methyl-imidadzolium chloride an interesting reaction medium for the conversion of cellulose and hemicellulose to monomer, e.g. monosaccharides, furfural, 5-hydroxymethylfurfural (HMF). In aqueous phase these conversions are limited by the accessibility of the carbohydrates within the complex biomass matrix.

In this work, it is demonstrated that acid catalyzed conversion of loblolly pine wood in 1-butyl-3-methyl-imidadzolium chloride occurs at much lower temperatures than in aqueous phase. At 120 °C the entire carbohydrate fraction of wood is converted into water soluble products within 2 h, while the lignin fraction remains as an insoluble residue. The water soluble products include polysaccharides, monosaccharides, furfural and HMF. The influence of the reaction temperature, water concentration and acid concentration are investigated. Under more severe conditions (higher temperature, higher acid concentration), solid degradation products, so-called humins, are formed by re-polymerization of the dissolved monomers.

Further insight in the reaction network involved in these transformations was obtained by studying the acid catalyzed conversion of model compounds in 1-butyl-3-methyl-imidadzolium chloride was studied.

639a Enhancement of Enzymatic Digestibility of Microcrystalline Cellulose by Treatment in Subcritical Water

Sandeep Kumar, Rajesh Gupta, Y. Y. Lee, and Ram B. Gupta. Chemical Engineering, Auburn University, Auburn, AL 36849-5127

Microcrystalline cellulose was treated with sub-critical water in a continuous flow reactor for enhancing its enzymatic reactivity with cellulase enzyme. Cellulose-water slurries were mixed with preheated water at a mixing junction and rapidly heated to the reaction temperature. The reaction product was immediately cooled to the ambient temperature using a heat exchanger. The resulting cellulose was characterized for its crystallinity and enzymatic hydrolyzability. Experiments were conducted at temperatures ranging from 200 to 315 °C, at 27.6 MPa, and reaction times of between 3.4 and 6.2 s in a continuous flow apparatus.

The samples were examined by degree of polymerization test, XRD, FTIR, and SEM analyses for assessing any transformation in the cellulose structure. The findings show that the degree of polymerization of cellulose treated at 315°C is decreased substantially, which helped in the enhancement of its enzymatic digestibility. SEM pictures of such samples show the development of cracks on the cellulose surface thus increasing the accessibility of enzymes. Nearly three folds increase in enzyme reactivity is observed after one hour of treatment with cellulase enzyme at loading of 3.5 FPU/g glucan, for cellulose samples treated at 315°C in sub-critical water. However, XRD and FTIR analysis results did not show any decrease in crystallinity index for all the samples.

Web Page: www.eng.auburn.edu/~gupta

639b A Scraped Surface Bio-Reactor for Batch Enzymatic Saccharification of Pretreated Corn Stover Slurries

Rajesh K. Dasari[1], Kyle W. Dunaway[2], and **R. Eric Berson**[2]. (1) BioEnergy International, Woburn, MA 01801, (2) Chemical Engineering, University of Louisville, Louisville, KY 40292

High solids processing of biomass slurries provides the following benefits: maximized product concentration in the fermentable sugar stream, reduced water usage, and reduced reactor size. However, high solids processing poses mixing and heat transfer problems above about 15% for pretreated corn stover solids. High solids slurries exhibit high viscosities and require high power consumption in conventional stirred tanks since they must be run at high rotational speeds to maintain proper mixing. An 8 liter scraped surface bio-reactor (SSBR) is employed here for enzymatic saccharification experiments in order to handle high solids loading and as a means for scale-up from lab-scale shake flasks. The scraping action of the blades keeps the reactor surface clear, which improves the heat transfer characteristics. The horizontal rotation of the shaft and blades provides mixing and prevents particle settling much more effectively than in conventional stirred tanks, even at very low rotational speeds. The reactor is designed to be easily scaleable to pilot plant or production scale. The role of the viscosity of biomass slurries in power consumption of the reactor is presented. The efficiency of the saccharification reaction is defined as sugar released per unit energy input. Tests were performed for solids loadings between 10% and 25% with results showing that the efficiency factor is the highest for 20% initial solids concentrations. This efficiency factor can be used to optimize future design and processing strategies.

639c Enzo-Rheology: Investigations of High-Solids Biomass Slurries for Bio-Refinery Applications

Jeffrey Knutsen and **Matthew W. Liberatore**. Department of Chemical Engineering, Colorado School of Mines, 1613 Illinois St, Golden, CO 80401

The enzymatic digestion of cellulosic biomasses, such as corn stover and switchgrass, to create biofuels, such as ethanol, is economically competitive only when working at high solids fractions. Therefore, the production of cellulosic ethanol involves working with high-solids slurries, which are notoriously difficult to transport due to their high viscosity, requiring large, powerful pumps. As a result, it is particularly important to understand the rheological properties of these slurries. In this work, we examine the rheology of high-solids biomass slurries as a function of the critical variables for biorefining. Acid hydrolyzed corn stover slurries are investigated and found to exhibit properties characteristic of soft solids, including an apparent yield stress and shear thinning behavior. Rheological properties were examined using vane-in-cup and parallel plate geometries. Special care is taken to account for or avoid wall slip in the rheological measurements. Shear thinning behavior and an apparent yield stress on the order of 1000 Pa were observed, and the Herschel-Buckley model was used to characterize the data. Rheological properties of the slurries were also measured during enzymatic hydrolysis. Interestingly, preliminary measurements suggest that viscosities do not decrease monotonically with the degree of cellulose conversion (digestion time). An attempt was made to correlate the slurries' rheological properties with the enzymatic hydrolysis of cellulose and the molecular interactions between the slurry particles. Thus, "enzo-rheology" combines enzyme kinetics, rheology and microscopy to capture the breakdown of fibrous cellulosic biomasses and the processability of the slurries.

Web Page: www.mines.edu/~mliberat/

639d Size Reduction Fundamentals for Biomass Conversion

J.Y. Zhu, USDA Forest Service, Forest Products Laboratory, One Gifford Pinchot Dr, Madison, WI 53716

Sieving method has been almost exclusively used for biomass size reduction characterization in the published biomass refining literature. This study demonstrates a methodology for properly characterize specific surface of biomass substrates through two dimensional measurement of each fibers of the substrate using a wet imaging technique. The methodology provides more information than the sieving method for properly characterize biomass substrate for biorefining. The image measured dimensions of individual fibers were used to estimate the substrate external surface based on a cylinder model. The substrate specific surface and mechanical refining energy consumption were then correlated to enzymatic hydrolysis glucose yield. Results indicated that the developed methodology is effective to differentiate various size-reduction and chemical pretreatment processes in terms of cellulose conversion efficiencies and size-reduction energy consumption. Disk milling-I that exposes cellulose is more effective than disk milling-II that exposes lignin in enzymatic hydrolysis. However, DM-I is more energy intensive than DM-II even taking the thermal energy into consideration. Both disking millings that produce fibers are more efficient in enzymatic hydrolysis than hammer milling that produce fiber bundles. Chemical pretreatment is not only increased cellulose conversion, but also reduced mechanical refining energy consumption. The present methodology correctly identified a proprietary pretreatment C as the most efficient pretreatment in terms glucose yield and refining energy consumption.

639e Conversion of Lignocellulosic Sugars to Triacylglycerols: A Renewable Diesel Feedstock

Guochang Zhang, Swalm School of Chemical Engineering, Mississippi State University, P.O.Box 9595, Mississippi State University, MS 39762, Dr. William T. French, Chemical Engineering, Mississippi State University, Starkville, MS 39759, Rafael Hernandez, Dave C. Swalm School of Chemical Engineering, Mississippi State University, 330 Swalm President's Circle Dr., Mississippi State, MS 39762, Earl G. Alley, Mississippi State Chemical Laboratory, Mississippi State University, Mississippi State, MS 39759, and Maria C. Paraschivescu, Department of Chemical Engineering, Mississippi State University, P.O.Box 9595, Mississippi State University, MS 39762.

Microbial conversion of lignocellulosic sugars to triacylglycerols, which could be further transesterified to make biodiesel, was investigated using the oleaginous yeast Rhodotorula glutinis (ATCC 15125). In the experiments, the yeast was continuously grown in a nitrogen-rich medium utilizing an artificial acid hydrolysate of lignocellulosic biomass switchgrass as the sole carbon and energy source. Once the culture had reached the stationary phase, the cells were harvested and transferred to a fresh nitrogen-free media containing artificial acid hydrolysate for lipid accumulation. The yeast grew very well in artificial acid hydrolysate mediums. Also in this study, R. glutinis was found to metabolize glucose first followed by xylose after the glucose was exhausted. The lipid accumulated in the yeast, determined by gravimetrical method, increased to 33.96% of dry cell mass while growing on 45 g/l of pure glucose, and 44.02% for using 45 g/l of pure xylose as sole carbon and energy source. When growing on artificial acid hydrolysate medium with a total sugar of 45 g/l in the second lipid accumulation phase, the lipid content of cell biomass was determined to be up to 29.76%. The majority contents of triglycerides were characterized by high temperature gas chromatography to be of C16~C20. This study demonstrated that it is feasible for oleaginous yeasts to convert lignocellulosic sugars to triglycerides: a renewable diesel feedstock.

639f Demonstration of a Viable Hydrogenase Assay for Syngas Fermentation to Ethanol: Quantitative Accounting for Competing Reactions, Enzyme Inhibition, and Diffusion Limitations

Bradley E. Skidmore, Jason M. Bray, Douglas R. Tree, and Randy S. Lewis. Chemical Engineering, Brigham Young University, 350 Clyde Building, Provo, UT 84602

Rising fuel prices, concern about long term crude oil supply, and increasing anthropogenic carbon dioxide emissions have increased motivation to find alternatives to traditional oil-derived transportation fuels. Among the most promising alternative fuels is ethanol, which in the United States is produced almost entirely from corn. The simple-sugar fermentation process that is currently used is inefficient in that it wastes much of the corn plant and costs almost as much energy to produce as is generated by its combustion.(Shapouri, 2004) A more promising alternative is to utilize the entire plant and expand the spectra of feedstocks that can be used for ethanol production into the non-food sector. Syngas (a mixture composed primarily of CO, CO_2, and H_2) is easily created by gasifying any carbon-based feedstock (e.g. switchgrass, municipal waste, forestry residues) and has been demonstrated to be able to be fermented by specialized microorganisms to make ethanol. (Datar, 2004)

As part of the fermentation process of converting syngas to ethanol, the hydrogenase enzyme(s) plays a critical role in converting H_2 to electrons for use in the metabolic process. If hydrogenase activity is diminished, CO is oxidized to CO_2 to provide the needed electrons. Thus, CO used in this aspect cannot be used to produce ethanol. Therefore, properly assessing the hydrogenase activity in the presence of syngas constituents is critical in understanding how to maximize ethanol productivity.

Many assays exist for determining the hydrogenase activity although numerous studies show varying results. One of the most prevalent methods is using an artificial electron acceptor (usually a color changing dye such as methyl or benzyl viologen) and a spectrophotometer to monitor the rate of hydrogen oxidation. Hydrogenase activity can vary depending upon the environmental conditions and the type of bacteria. However, the activity can be incorrectly measured as a result of experimental protocols. Previous research has shown that oxygen scavengers used to keep the assay anoxic can result in an additional reaction of the scavenger with the electron acceptor, leading to an overestimated measurement of the hydrogenase activity (Van Dijk, 1979). Additionally, it has been shown that hydrogenase activity in some assays may be limited by the rate at which H_2 can diffuse to the surface of the cells (Tatsumi, 2000). Finally, CO (a component of syngas) is a known inhibitor of hydrogenase activity. (Chen, 1974)

As a result of potential side reactions interfering with an accurate hydrogenase assay, this work quantified the reaction rate between one common oxygen scavenger (dithiothreitol) and one common electron acceptor (benzyl viologen). Also, rate data from CO inhibited hydrogenase assays was fit to a modified Michaelis-Menton equation and an inhibition constant (K_i) was obtained. Finally, diffusional analysis was applied to show that H_2 in the assay solution cannot be replenished by H_2 in the headspace during the analysis time. The diffusional analysis, inhibition constant, and knowledge of the side reaction rate provided the background to develop a viable hydrogenase activity assay. This work will describe the hydrogenase assay and demonstrate how to accurately account for diffusional effects, inhibition, and side reactions.

References

Chen J, Blanchard D. Isolation and properties of a unidirectional H_2-oxidizing hydrogenase from the strictly anaerobic N_2-fixing bacterium Clostridium pasteurianum W5. Biochemical and Biophysical Research Communications 1978;84:1144-1150.

Datar R, Shenkman R, Cateni B, Huhnke R, Lewis R. Fermentation of bio-mass generated producer gas to ethanol. Biotechnology and Bioengineering 2004;86:587-594.

Shapouri H, Duffield J, Mcaloon A. The 2001 net energy balance of corn-ethanol. Proceedings of the Conference on Agriculture as a Producer and Consumer of Energy, Arlington, VA., June 24-25, 2004.

Tatsumi H, Kano K, Ikeda T. Kinetic Analysis of Fast Hydrogenase Reaction of Desulfovibrio vulgaris Cells in the Presence of Exogenous Electron Acceptors. J. Phys. Chem. B 2000;104:12079-12083.

Van Dijk C, Mayhew S, Grande H, Veeger C. Purification and Properties of Hydrogenase from Megasphaera elsdenii. Eur. J. Biochem 1979;102:317-330.

High Yielding Biorefining using Organosolv Processing

E. Kendall Pye, Ph.D., William B. Armiger, Ph.D., Stephen R. Winner, Lignol Innovations Inc., Berwyn, PA, USA,

Alex Berlin, Ph.D., Lignol Innovations Ltd., Burnaby, BC, CANADA

Introduction

Various economic, environmental and political forces have combined to create an unprecedented change in the world economic order in less than a single decade. Rapidly expanding economies in the world's most populous nations have created a large consumer-oriented middle class resulting in an increased demand for energy, higher quality foods and material resources. These factors together with mounting evidence that sources of easily and cheaply obtainable oil are declining, has led to rapidly escalating crude oil prices, which have doubled during the past year. Within the developed countries other pressures, such as supply concerns, national security, balance of trade issues and environmental problems including global warming, have led to recent legislation and the commercial development of alternative forms of energy, especially renewable energy. Various forms of renewable energy, including wind, solar, tidal and biomass-derived liquid transportation fuels, are now undergoing rapid commercialization. The most immediate problem facing the US general population is the rapidly escalating price of liquid transportation fuels, home heating fuels and secondarily the price of oil and natural gas-derived chemicals and materials. These are having a highly inflationary effect on prices in general and food prices in particular.

The New Industry of Biorefining

- **Numerous concerns are now stimulating the drive for fuels and chemicals from renewable lignocellulosic materials**
  - **Cost of crude oil; impact on liquid fuels and chemicals prices and on national economies**
  - **The "peak oil" scenario; ever increasing demand and prices**
  - **Climate change from fossil carbon greenhouse gases**
  - **Energy security for oil importing nations**
  - **Balance of trade problems for oil importing countries**

The New Industry of Biorefining

As a consequence of these factors there has been a dramatic expansion of corn-based ethanol facilities in the US. By the end of this year, approximately 170 such facilities should be in production with an annual production capacity greater than 9 billion gallons of ethanol. Ultimately, according to the 2007 US Energy Independence and Security Act a total of 15 billion gallons of corn-based ethanol will be required by 2022. However, that same act mandates that there should be an additional 21 billion gallons of "Advanced Biofuels" of which

16 billion gallons should be from lignocellulosic sources. This calls for an enormous construction effort, which will introduce a totally new national industry of biorefining of cellulosic materials such as wood, agricultural residues, deliberately grown energy crops, and the like.

Biorefinery Concepts

Lignocellulosic biomass, the structural components of plants such as tree trunks and cereal straw, is now the primary focus of the developing biorefining industry because it represents the lowest cost and largest potential volume of any plant material on the planet. Vast quantities of woody agricultural residues, such as straw and corn stover, are presently either left in the field or incinerated for no value. Such materials are composed primarily of three biochemical polymers. The most abundant one is cellulose, a homopolymer of the sugar D-glucose, linked by a β, 1-4 linkage. A lesser polymer is hemicellulos*e*, a complex polymer consisting of various hexose and pentose sugars, together with acetyl groups and uronic acids. The third polymeric component is lignin that is made up of variously modified phenylpropane moieties linked by numerous types of linkages including carbon-carbon bonds and aryl-alkyl ether linkages. In addition to these primary components of woody biomass, there are minor components including various lipophylic compounds generally grouped under the heading of "extractives" that include resin acids, terpenes, fatty acids and phytosterols. Many minor components of woody biomass have been the source of valuable chemicals that have found uses in medicine (e.g. salicylic acid, β-sitostanol, etc.), in food (e.g. arabinogalactan), and in materials (e.g. turpentine).

Biorefineries capable of processing these relatively recalcitrant lignocellulosic feedstocks can use many different technologies. A conceptually relatively simple one is the gasification of biomass to create syngas, which can then be converted into a somewhat limited range of organic chemicals by chemical or biological means. Pyrolysis is another thermochemical process that creates a range of gaseous, liquid and solid products. Like gasification, pyrolysis has the advantage of being relatively insensitive to the type of feedstock, but both of these thermochemical processes significantly degrade the complex chemical components in lignocellulosic biomass, many of which have potentially high commercial value in their original form.

Other biorefinery processes utilize strong or dilute acids, steam or alkali to open up the compact cellular structures in woody biomass, thus exposing the fiber wall polysaccharides, cellulose and hemicellulose, to enzymatic attack. Most of these processes target the production of hexose and pentose sugars derived from the hydrolysis of cellulose and hemicellulose for conversion into fermentation products such as ethanol, butanol, lactic acid, etc. The remaining impure biomass components, such as lignin, rosin and resin extractives, etc. are then mostly used as low value solid fuel to power the process.

Organosolv Biorefineries

Organosolv technology, as used in the ethanol-based Lignol Modified Organosolv Biorefinery Process, allows for the fractionation of lignocellulosic biomass into a series of valuable chemical products, which have a combined commercial value that greatly exceeds the value of a variety of fermentation products made from just the carbohydrate fractions of woody biomass alone. The advantages of this value-retention strategy are numerous. With greater revenues being obtained from each unit of feedstock, such plants require smaller amounts of resource to be profitable, thus avoiding the need for large harvesting areas and opening up many more suitable locations for biorefineries. This has the added advantage that profitable plants can be smaller, thus requiring a lower total capital investment, which is a considerable advantage for an early stage technology. Furthermore, the revenues of such a plant are less dependent on the potentially-variable market values of any one single product, such as ethanol. The process treats woody biomass, - softwoods, hardwoods, agricultural residues, deliberately grown energy crops, etc. – with an aqueous ethanol liquor at elevated temperatures. The precise conditions are dictated by the nature of the feedstock, the desired product streams and process energy considerations.

The lignin and hemicellulose in the biomass are partially hydrolyzed and dissolve into the liquor, while the hydrophobic "extractives" also dissolve in the hot organosolv liquor. Some additional chemistry occurs in the process, such as the conversion of some pentose sugars from the hemicellulose into furfural, but generally, most of the fine chemical structures originally in the biomass are preserved. Following the modified organosolv stage, a solid fraction consisting mostly of cellulose with minor amounts of residual lignin and hemicellulose is ready to be converted by enzymatic hydrolysis into a sugars stream useful for various fermentation processes. Because most of the lignin has been removed, this cellulose-rich fiber mass is more readily hydrolyzed by enzymes than are equivalent streams from other biorefinery processes. This "pulp" stream can also be processed further into a number of commercially-useful chemicals and materials, such as cellulose derivatives and powders. The liquor from the organosolv stage is processed to recover separate streams of high-purity lignin (HP-L™ lignin) for use in resins and a large number of other applications, as well as furfural, acetic acid from the hemicellulose, wood "extractives", pentose and hexose sugars and oligosaccharide products. The ethanol used in the process is recovered by distillation and recycled back to the process. Energy for the process, - steam and electricity only – can be provided from co-generation facilities burning bark and forestry residuals. No fossil carbon fuel is required for the process. The net water consumption of the process is negligible.

The unit processes that make up the Lignol Biorefinery are combinations of well-known processes that currently exist in either the chemicals, biochemicals or pulp and paper industry. Some, but very little new process equipment is required, assuring a high degree of confidence in its operability and performance. A large 70 tonne of feedstock per day demonstration plant utilizing a substantial portion of this technology has been operated for many years in the past. The multi-product, value-retention approach to biorefining has numerous technical and economic advantages, which will be discussed.

The Lignol Biorefinery technology, which is able to process softwood, hardwood and annual fibers, consists of two principle steps (Figure 1). First is a modified solvent extraction stage that treats biomass with an aqueous alcohol liquor at elevated temperatures and pressures for a specified time. This treatment converts much of the large lignin polymers in the woody biomass into smaller molecular weight fragments, which dissolve in the hot alcohol-based liquor. These fragments still retain most of the chemical structure and properties of the original lignin, but unlike lignin extracted from wood in the kraft and sulfite pulping industry, they do not have sulfur introduced into their chemical structure. Thus the lignin extracted from biomass by the Lignol technology still retains many valuable properties above and beyond its aromatic content and finds high value commercial markets.

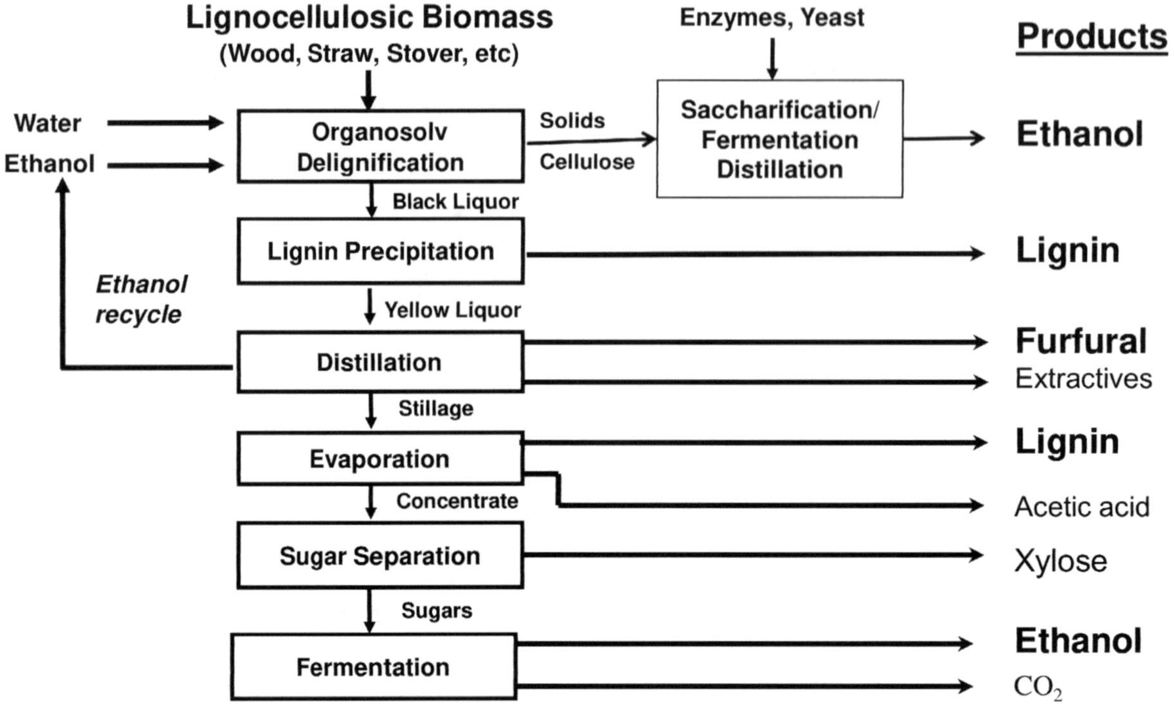

Figure 1. The Lignol Biorefinery Process

A number of other important reactions occur during this first extraction stage of the Lignol Biorefinery process. Acetic acid, furfural and various hexose and pentose sugars arise from reactions involving hemicellulose. Additionally, most of the lipophylic components of the biomass, generally called "extractives", appear in the hot liquor. The products of this first stage, therefore, are a cellulose-rich fiber and an aqueous alcohol liquor containing the extracted materials. This liquor is then treated in a series of unit processes to recover the ethanol; the industrial chemical furfural; a valuable extractives fraction; additional lignin fractions and the hemicellulose hexose sugars, which can be fermented to produce ethanol. Acetic acid, a widely used industrial chemical, can also be recovered from the liquor.

The second stage of the Lignol Biorefinery process is the conversion of the cellulose in the fibers to glucose, a process called saccharification, and then the fermentation of this glucose to produce ethanol. Due to the relative purity of the cellulosic fiber fraction that results from this process, Lignol uses a cellulase enzyme complex to perform the saccharification. This results in very high yields and conversion rates that cannot be achieved with acid catalyzed saccharification. The yeast that is employed in the fermentation has been carefully selected by Lignol to allow rapid ethanol production from both the cellulose-derived glucose as well as certain hexose sugars derived from the hemicellulose.

As can be seen, the Lignol Biorefinery creates high value from the biomass feedstock by producing multiple co-products together with high yields of fermentable sugars that are readily converted to fuel ethanol, or other "sugar platform" chemicals. This is achieved to a great extent by preserving the structure and then recovering most of the chemicals that nature created in the biomass. Many other biorefinery processes currently under development, utilize only the sugars obtained from the cellulose and sometimes the hemicellulose, and then burn the remainder to create steam and power for the process. This provides something less than solid fuel value to the plant. There are many other materials, less amenable to biorefining, such as bark, that can provide fuel to a biorefinery.

The Lignol Biorefinery process has numerous advantages that result mostly from its production of multiple co-products. The revenue from these additional products creates an economic advantage that can be exploited to allow the technology to be viable in relatively small plants. This opens up the prospect of locating these biorefineries in large sawmills to use the wood residues created at these operations, or placing them within pulp mills where excess wood chips could be made available. Certainly, this technology holds the promise of creating significant value from fuels and chemicals derived from lignocellulosic biomass and should be a leading technology in the new rapidly developing biorefining industry.

Pure Lignin – A Product of Organosolv Biorefineries

As one of the three major polymers always present in woody biomass, lignin represents a considerable proportion of the structural components of plants. It can account for approximately 10-12% of the aerial portion of some short annual plants and up to 30% or more for some coniferous trees. As such it is claimed to be the second most abundant organic chemical on earth [and is therefore a major renewable chemical that should not be overlooked, either in volume or value. Significant commercial markets, totaling over one million tons per year, already exist for lignins that have been recovered from chemical pulp mills . The value of lignin in these markets is generally an order of magnitude higher than its fuel value. With the anticipated future construction of biorefineries processing lignocellulosic feedstocks, the amount of lignin potentially available for marketing for its chemical value, rather than its fuel value, is likely to be enormous. Furthermore, lignin from some types of biorefineries will most likely have better performance characteristics and be of greater commercial value for its chemical properties than lignins from existing chemical pulping operations. The availability of these "new" lignins with enhanced physical and chemical properties will no doubt stimulate major new markets for this renewable material. However, it is likely that these materials will find their initial markets in the same sectors as the current lignin products. A review of the manufacturing, properties and present markets for lignins from pulping operations is therefore instructive in assessing the potential economic value of lignin that might be recovered from biorefineries processing lignocellulosic materials.

A simple examination of the chemical structure of native lignin [shows some interesting similarities to the structure of phenol-formaldehyde resins. Lignin consists of a three-dimensional amorphous matrix of cross-linked phenylpropanoid units. These residues, depending on their botanical source, may have, to a greater or lesser extent, substitution of methoxy groups on the ortho positions to the hydroxyl group on the aromatic ring. Lignin from annual plants, with a high proportion of ρ-hydroxy phenylpropane residues, will have a significantly lesser degree of o-substitution, while lignins from hardwoods, with a high content of syringyl moeties, will have a high proportion of o-substituents on the aromatic ring. Softwood lignin with a high content of guaiacyl groups (more than 90%) and some p-hydroxy phenylpropane moieties will have a greater degree of methoxy substitution.

Both native lignin and phenol-formaldehyde resins are polyphenolic materials consisting of phenolic rings linked together by short aliphatic chains. However, the form of these linkages in lignin are much more diverse and different from the linkages in PF resins. In lignin many of the cross-linking groups are aryl-alkyl ethers, but there are numerous carbon-carbon links also. Furthermore, lignin has a significant number of functional groups, such as aliphatic hydroxyls, on the propane residues, a feature that is not shared with phenolic resins. Nevertheless, the structural similarities are close enough to suggest a strong compatibility between these two materials and even opportunities for chemical interaction between them, with or without added cross-linking agents. Lignosulphonates and thiolignins have been marketed for incorporation into phenolic resins for some time, but they do not appear to have met with significant acceptance by the resin industry.

Some of the problems experienced by current lignins in this market may be overcome with biorefinery lignins, but not by all forms of biorefinery lignins. Organosolv processing of woody biomass can readily yield a non-derivitized lignin having high purity, significant chemical reactivity, low polydispersity and low molecular weight. Organosolv lignins from hardwood produced in tonnage quantities by the Alcell process in Miramichi, Canada, in the 1990's has been readily substituted into phenolic resins and used successfully on a commercial basis. Commercial applications in which it was used included in phenolic resins used in OSB manufacture and the production of friction materials. Numerous other applications where organosolv lignin was being substituted in part or in whole for phenolic resins were also under development at that time, such as in saturating resins for laminates production, stiffening agents for container boards and in rubber processing additives.

The manner of substituting lignin for phenol-formaldehyde resins can vary from a) a simple blending of dry powder lignin with dry powder phenolic resin, to b) the use of organosolv lignin as a primary phenolic component during the manufacture of the resin. In both cases the lignin represents the substitution of a renewable material for non-renewable chemicals produced from fossil carbon.

The advantages to the users and suppliers of phenolic resins of partial or complete substitution of biorefinery lignin are numerous. They include the potential of lower cost raw material and the opportunity to receive valuable carbon credits since phenolic resins are produced from phenol, now almost totally produce from crude oil, and formaldehyde, which is a product of natural gas. In addition, recent studies have shown that the incorporation of organosolv lignin into PF resins that are used in the manufacture of OSB will reduce the emissions of formaldehyde from the press during manufacture. This is a major advantage to manufacturers because these emissions of a known carcinogen are being closely monitored and regulated by various governments. A added benefit to OSB manufacturers is that it has been shown that the substitution of up to 35% of PF resin with an equal weight of organosolv lignin will provide substantial improvements to the final board properties, especially in reduced swell of the wet board and a higher modulus of rupture following the boil test.

664b Mixtures Effects and Kinetic Characterization of Timber and Switchgrass Diluted Acid Hydrolysis

Jill R. Jensen[1], Juan Morinelly[1], Anwar Aglan[1], Adam Mix[1], and David R. Shonnard[2]. (1) Chemical Engineering, Michigan Technological University, 1400 Townsend Dr., Houghton, MI 49931, (2) Department of Chemical Engineering, Michigan Technological University, 1400 Townsend Dr., Houghton, MI 49931

In recent years, growing attention has been devoted to the use of lignocellulosic biomass as a feedstock to produce renewable carbohydrates as a source of energy products, including liquid alternatives to fossil fuels. Currently, research is driven by the need to reduce the cost of biomass-ethanol production. One of the prominent methods is to thermochemically hydrolyze (pretreat) the biomass and subsequently, enzymatically hydrolyze the pretreated material to fermentable sugars that can then be converted to ethanol using specialized microorganisms. A key goal of pretreatment is to produce fermentable sugars from hemicellulose and enhance enzymatic conversion of the cellulose fraction, and, hopefully, obtain a higher ethanol yield. Most previous studies on biomass pretreatment have involved pure species, but mixtures of biomass are likely to be feedstocks for agricultural and forest-based biorefineries. The primary goal of this research was to obtain kinetic detailed data for dilute acid hydrolysis from mixtures of several timber species from the Upper Midwest region of the United States and switchgrass, and to determine whether there were any synergistic or antagonistic effects due to the simultaneous pretreatment of mixtures.

Dilute acid-catalyzed hydrolysis (pretreatment) of biomass, including aspen, basswood, balsam, red maple, and switchgrass was studied. Initial experiments were conducted to pretreat each pure species. Following was an experimental design that pretreated 50/50 weight percent mixtures of each possible combination of the five species in duplicate. Overall, there were 25 experiments, 5 pure species and 20 mixtures conducted in a 1-L well-mixed batch reactor (Parr Instruments, Model 4570). 25g of biomass and 500mL of 0.5% sulfuric acid were added to a 1-L glass liner, and then put into the reactor and heated from room temperature at a rate of about 3 degrees C/min. During the experiment, 5 ml samples were taken starting at 130 degrees Celsius at 3 min intervals until reaching the temperature setpoint of 175 degrees Celsius, followed by 4 samples after achieving the setpoint temperature. The collected samples were then cooled in an ice bath to stop the reaction. The cooled samples were filtered using a 0.2 micron MILLIPORE membrane filter to remove suspended solids and analyzed using High Performance Liquid Chromatography (HPLC) with a Bio-Rad Aminex HPX-87P column, and refractive index detection to measure monomeric sugars and diode array detection to measure the main degradation byproduct, furfural. An additional acid hydrolysis was completed to convert any oligomeric sugars into monomers which were also detected by HPLC for a total sugar analysis.

A first order reaction model that takes into consideration simultaneous formation and degradation of xylose was used and the kinetic parameters such as activation energy and pre-exponential constant were obtained for the pure species. Excellent agreement between the model and pure species experimental data was obtained. These parameters were then used to predict the mixtures yields, also with excellent agreement between experimental data and the predicted model. No significant synergistic or antagonistic effects were observed from pretreating mixtures of the five species studied compared to single species results. These results confirm that dilute acid hydrolysis of woody biomass species in mixtures is expected to proceed as if the species existed in the reactor alone (Jensen et al., 2008). This outcome could suggest strategies for maximizing sugar yields from mixtures based on knowledge of how pure species behave during dilute acid hydrolysis.

References

Jensen J, Morinelly J, Aglan A, Mix A, Shonnard DR. Kinetic characterization of biomass dilute sulfuric acid hydrolysis: Mixtures of hardwoods, softwood, and switchgrass. AIChE J. 2008; 54:1637-1645.

664c The Effects of Impregnation Techniques on Pretreatment Performance for High Dry Matter Dilute Sulfuric Acid Processes

Noah D. Weiss, Nicholas J. Nagle, Melvin P Tucker, Erik Kuhn, and Richard T Elander. Bioprocessing R&D, National Renewable Energy Laboratory, 1617 Cole Bolevard, Golden, CO 80401

Pretreatment is an integral part of the conversion process of cellulosic feedstocks to biofuels via biochemical process routes. However, it is also one of the most costly and energy intensive steps of the process. Many pretreatments have been demonstrated effectively in the laboratory under conditions which would not be feasible at commercial scale. It is therefore important to demonstrate scaling of pretreatment processes as part of a rigorous evaluation and demonstration of their commercial potential. High dry matter (DM) content is an important factor in large scale pretreatment, but can present significant issues in scale up and process performance.

One of the key process parameters for some catalyst based pretreatments is the handling of biomass before pretreatment and the impregnation of the biomass with the catalyst. Especially when operating at high solids concentrations, catalyst impregnation can greatly impact the efficacy of a given pretreatment. In this study we evaluated a wide range of process parameters related to catalyst impregnation thought to impact pretreatment results of dilute sulfuric acid pretreatment. Corn stover was impregnated under various preprocessing scenarios and pretreated in a steam digester under low and high pretreatment severity conditions. This was carried out using a Placket-Burman experimental design meant to screen many processing scenarios and factors for significance. Factors investigated included impregnation time and temperature, spraying versus soaking impregnation methods, particle size, vacuum application and passage through a compression screw feeder. All samples were pretreated at an initial dry matter content of 45% in a batch steam digester. Initial results pointed to the method of catalyst impregnation as the most significant factor, along with catalyst loading during impregnation. A follow up study was carried out to investigate the amount of excess acidified liquid necessary to apply to the biomass to maximize xylose yields, by varying the impregnation solids concentration and sulfuric acid loading. Results show the impacts of these process variations on xylose yield and enzymatic digestibility, while providing information on optimal regions for the operation of catalyst impregnation processes. The impact of these results on industrial scale impregnation scenarios will be discussed along with their implications for future processing scenarios for continuous reactors.

664d Integrated High Throughput Pretreatment and Enzymatic Hydrolysis In 96 Well Plates

Michael Studer[1], Jaclyn D. DeMartini[2], Heather L. McKenzie[2], and Charles E. Wyman[2]. (1) CE-Cert, University of California Riverside, 1084 Columbia Avenue, Riverside, CA 92507, (2) Chemical and Environmental Engineering, University of California Riverside, CE-CERT, 1084 Columbia Avenue, Riverside, CA 92507

The challenge of converting cellulosic biomass to sugars dominates the cost of biological processing to ethanol and other products. The BioEnergy Science Center (BESC), led by Oak Ridge National Laboratory, is addressing this challenge by advancing plants to reduce recalcitrance and new biocatalysts to improve deconstruction. Systems biology in conjunction with new high-throughput (HTP) analytical and computational technologies will be instrumental to this approach. In order to enable the screening of the many different natural and genetically modified biomass types and biocatalysts and identify those with enhanced features for biological conversion, a new HTP tool for pretreatment and enzymatic hydrolysis system is being developed. In our approach, pretreatment with dilute acid or just water is integrated with enzymatic hydrolysis in the same 96 well plate. Each well is loaded with 2 – 5 mg of milled biomass that is suspended in water or dilute acid to a final slurry concentration of 1 - 2% w/w. After loading, the plate is sealed and rapidly heated to the target pretreatment temperature in a steam chamber and held there for the desired reaction time. The pretreatment reaction is then quenched by flooding the steam chamber with cold water. Samples which were pretreated with dilute acid are neutralized in situ, and citric acid buffer, sodium azide, and enzymes (cellulase and xylanase) are directly added to each well, with the enzyme loading based on the original glucan and xylan content of the raw biomass. Next, the reactor is re-sealed and placed in an incubator shaker at 50°C for 72 hours, and the hydrolyzate sugar concentration in each well is analyzed using a redox enzyme-mediated spectroscopic assay. Applications of this integrated approach will be presented for poplar wood and switchgrass.

664e Efficient Sugar Release from Common Reed Pretreated by Cellulose-Solvent-Based Lignocellulose Fractionation at Lower Cellulase Loadings

Noppadon Sathitsuksanoh, Zhiguang Zhu, and Y-H. Percival Zhang. Biological Systems Engineering, Virginia Polytechnic Institute and State University, 200 Seitz Hall (0303), Blacksburg, VA 24061

Current cellulosic ethanol production is hindered by the high processing costs, capital investment, and relatively low sugar yields. Efficient conversion of biomass to sugars is the last obstacle to emerging cellulosic ethanol industries. Common reed is a very aggressive, robust, and densely growing grass that is posting ecological threat by crowding out native plants. Common reed has a very high biomass productivity per acre per year. Utilization of common reed as a bioenergy feedstock will bring 2-fold benefits, a control over this invasive plant and an increase in energy security. A new cellulose-solvent-based lignocellulose fractionation (CSLF) technology has been developed for isolating lignocellulose components based on their different solubilities in different solvents (Zhang et al. Biotechnol. Bioeng. 2007. 97: 214). The optimal pretreatment conditions for common reed were found to be 84% H_3PO_4 at 50°C for 60 minutes. To reduce enzyme utilization, we further investigated to decrease the enzyme loading from 15 FPU of cellulase per gram of glucan to lower levels without significant reduction in sugar yields. Scanning electron microscopy (SEM) images clearly show that CSLF can completely destruct fibril structure of common reed, resulting in a drastic increase in cellulose accessibility to cellulase.

Understanding Lignin Fractions of Biomass during Dilute Acid Pretreatment Using a Flow through Reactor

Yun Ji, Melvin Tucker, Rui Katahira, David Johnson, William Michener, Nick Nagle and Todd Vinzant
National Renewable Energy Laboratory,
1617 Cold Blvd, Golden, CO, 80401

Considerable research has been conducted into understanding the distribution of lignin in pretreated biomass and the effects of this distribution on enzymatic hydrolysis for bioethanol production. Enzyme accessibility is postulated to be one of the key factors affecting enzymatic conversion of pretreated biomass into fermentable sugars, with the lignin-hemicellulosic sheath seen as a barrier to enzymatic hydrolysis. It has been shown that thermal chemical dilute acid pretreatments of biomass disrupt this lignin-hemicellulosic matrix in such a way as to allow hydrolysis of hemicellulose into soluble sugars, and to coalesce lignin away from the cellulose microfibril thereby increasing localized enzymatic accessibility. Published results suggest that lignin redeposition could be a problem for enzymatic hydrolysis in batch reactor designs (Selig et al 2007). In this study, lignin distribution and morphology during dilute-acid pretreatment of corn stover are investigated using a flow through reactor at different reaction temperatures (150°, 170°C and 190°C) and different acid concentrations (0, 0.5 and 0.8 wt% H_2SO_4). Pretreatments using the flow-through reactor system result in continuous removal of the solublized hemicellulose and lignin with the flowing liquid. The lignin is collected with the effluent fractions and precipitates upon cooling. Representative samples of the residual solids were digested with spezyme and xylanase enzymes to determine the effects of pretreatments. In addition, the residual solids were washed with a various dissociation agents to remove the surface associated lignin in an attempt to understand the lignin/biomass interaction and to account for the mass of the lignin that is involved with the re-deposition phenomena. The washed solids were then exhaustively digested enzymatically and the residual lignin analyzed. Three lignin fractions (acid soluble lignin from liquid, acid insoluble lignin from liquid and residual lignin in the pretreated solid) were collected and characterized using different analytical methods including solid state NMR, FTIR, and HPLC-MS for detailed compositional analysis.

The results show that thermal /chemical pretreatment (dilute acid) hydrolyzed most part of xylan in the biomass. At the same time, lignin is also somewhat depolymerized and becomes less entrained with the cell wall matrix. It is clear that most of the lignin remains coalesced and trapped in the residual solids (85-95%). However, as the native lignin is hydrolyzed, smaller molecular weight fractions seem to migrate from their original location to the point where a small population (5-15%) makes its way into the aqueous phase of the flowing pretreatment liquor. A small portion of the lignin is redeposited on the cooled cell wall surface. The results give an understanding of the effects of lignin mobility due to pretreatment on enzymatic digestibility with application to further process development of commercial bioethanol plants.

This work was supported by the U.S. Department of Energy under contract number DE-AC36-99GO10337 with the National Renewable Energy Laboratory.

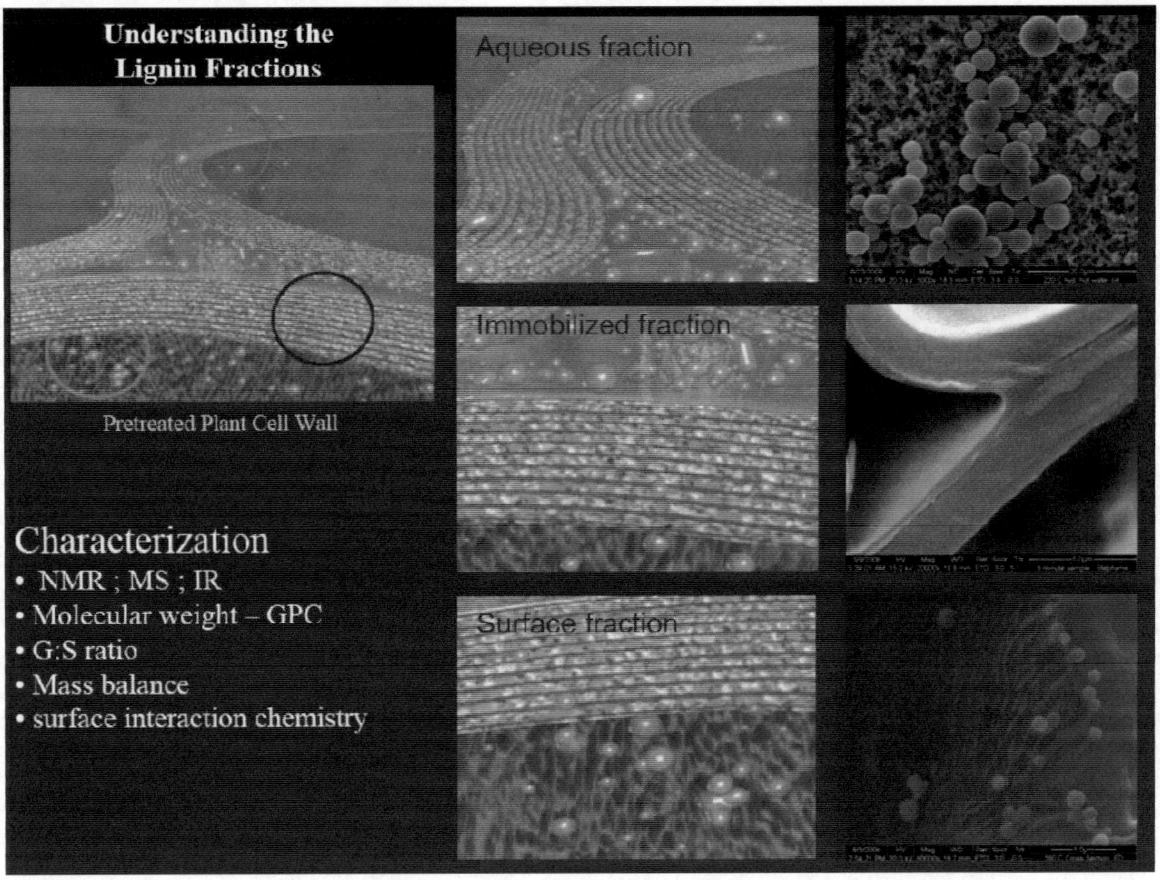
Understanding the Lignin Fractions
Pretreated Plant Cell Wall
Characterization
• NMR ; MS ; IR
• Molecular weight – GPC
• G:S ratio
• Mass balance
• surface interaction chemistry
Aqueous fraction
Immobilized fraction
Surface fraction

689a Novel Reactor Design for the Production of Biodiesel from Free Fatty Acids

Richard A. Cairncross, Cory Melick, and Minhazuddin Mohammed. Chemical & Biological Engineering, Drexel University, 3141 Chestnut St, Philadelphia, PA 19104

A semi-batch bubble column reactor has been developed for biodiesel production oils containing a high free fatty acid content. An acid catalyst was used to avoid soap formation because soap formation inhibits conversion and makes purification difficult. The reactor is operated at ambient pressure and temperatures above the boiling point of methanol. A major advantage of the bubble column design is continuous removal of byproduct water which enables overcoming equilibrium limitations of the reversible esterification reaction. A continuous reactor model is being developed to capture the kinetics of the esterification reaction and the effect mass transfer of the alcohol and water between the vapor and the oil phases. The reactor model will be validated based on a set of parametric studies consisting of alcohol flow rate; temperature and alcohol feed quality. Parametric studies were performed in a semi-batch lab scale to test the concept before the design and construction of a continuous reactor. The biodiesel conversion profile was obtained through titration of samples collected at regular intervals. The reactor is robust and efficient with conversion greater than 95% achieved within 60 minutes at 120°C reactor temperature. The bubble column design is robust for a variety of feedstocks. It achieves greater than 95% conversion with pure triglycerides, pure free fatty acid, and blends. The conversion is also unaffected by the presence of water (up to 5% by volume) in the methanol feed or using pure ethanol instead. It was also found that agitation does not effect the overall conversion, but does affect the rate. The results from these studies are being used to design a continuous counter-current reactor.

689b Optimization of the Synthesis of Biodiesel Via Ultrasound-Enhanced Homogeneous and Solid Acid Catalyzed Processes Using a Novel Reactor

Yusuf G. Adewuyi and Naresh N. Mahamuni. Chemical Engineering, North Carolina Agricultural & Technical State University, 1601 East Market Street, Greensboro, NC 27411

The recent attempts in the applications of ultrasonic and hydrodynamic cavitation for biodiesel synthesis present an urgent need to optimize the process to reduce cost of operation and make it economically attractive for large-scale applications. Biodiesel is an oxygenated diesel fuel consisting of mono alkyl esters of long chain fatty acids derived from renewable lipid feedstocks such as vegetable oils or animal fats, made up of triglycerides (TGs) and free fatty acids (FFAs) and is produced via the chemical process of transesterification. Biodiesel is not only biodegradable but it is also free of sulfur and aromatics. Biodiesel is produced by reacting the lipid feedstocks with an alcohol, such as methanol or ethanol in the presence of a catalyst (sodium or potassium hydroxide, or sodium methoxide) to yield methyl esters (biodiesel) and byproduct glycerin used to make soaps and cosmetics. This reaction does not tolerate the presence of water or fatty acids and the use of acids such as H2SO4, H3PO4, and HCl pose environmental and corrosion problems. The use of solid acids such as zeolites or zirconium sulfate would eliminate the need to use liquid acids or bases, the subsequent corrosion of process equipment and soap formation, and would make catalyst separation from products trivial and permit the use of continuous rather than batch processing. Cavitation is the formation, growth and implosive collapse of gas- or vapor-filled microbubbles and can be induced acoustically or hydrodynamically in a body of liquid. The collapse of these bubbles leads to local transient high temperatures (• 5000 K) and pressures (• 1000 atm), resulting in the generation of highly reactive species including hydroxyl (•OH), hydrogen (H•) and hydroperoxyl (HO2•) radicals, and hydrogen peroxide. Cavitation effects also increase the mass transfer and accelerates the reaction rate and yield of esters. However, little fundamental research has been carried out to date on the production of biodiesel via ultrasound enhanced homogeneous and heterogenous catalysis of esterificationa nd transesterification. Also, to aid the cost-effectiveness and commercialization of cavitation-enhanced bidiesel processes, the understanding of the effects of various ultrasonic parameters (e.g., frequency, intensity, etc) and other process parameters need to be understood and quantified. This talk will present results of our studies using a novel multi-frequency reactor to optimize and homogeneous and heterogeneous production of biodiesel.

689c Renewable Petroleum™ Technology: Production of Fit for Purpose Renewable Fuels

Scott A. Frykman, Department of Bioprocessing, LS9, Inc., 100 Kimball Way, South San Francisco, CA 94080

Current transportation fuels are predominantly derived from petroleum feedstocks and are a complex mixture of refined hydrocarbons. Petroleum, on which modern day society was built and is now dependent, is a diminishing resource with increasing environmental, political, and economic disadvantages. An ideal alternative would be chemically similar to petroleum, allowing broad and rapid adoption, derived from renewable resources, scalable to support current and future demands, domestically derived, and cost competitive. To meet this need, LS9 has developed a molecular tool box and a suite of industrial microorganisms that enable the selective production of a diverse portfolio of hydrocarbon products and derivatives. This talk will cover an introduction to the technology, an overview of some of the unique scientific challenges encountered, and the development of a scalable production process to enable the economic manufacture of these fuels in the near future.

PRODUCTION OF GREEN DIESEL BY HYDROCRACKING OF CANOLA OIL ON Ni-Mo/γ- Al_2O_3 AND Pt-ZEOLITIC BASED CATALYSTS

Rogelio Sotelo-Boyás[a], Yanyong Liu[b], and Tomoaki Minowa[b,*]
[a]*Instituto Politécnico Nacional, México* ESIQIE, México D.F. 07738, México
[b]*Biomass Technology Research Center – National Institute of Advanced Industrial Science and Technology, Chugoku,2-2-2 Hirosuehiro, Kure, Hiroshima 737-019, Japan*

* Corresponding author

Abstract

As an alternative way to produce biodiesel, the hydrocracking of canola oil was studied on different types of catalysts. Experiments were carried out in a batch reactor over a temperature range of 300 to 400°C, initial hydrogen pressure of 5 to 11MPa, and reaction times 1 to 6 hours. Three types of bifunctional catalysts were developed and used: Pt/H-Y zeolite, Pt/HZSM-5 and sulfided NiMo/γ- Al_2O_3. At the conditions studied, the hydrocracking on Ni-Mo/γ- Al_2O_3 catalyst gave the highest conversion to liquid hydrocarbons in the boiling range of diesel fraction. The Pt-zeolite catalysts revealed a strong catalytic activity for both cracking and hydrogenation reactions, and therefore initial high hydrogen pressures were found to be required to reach a good conversion. The time of reaction was limited in observance of the rates of reaction for both hydrogenation and cracking reactions, which strongly depended on the type of catalyst. In most cases, the GC analysis of the gaseous phase revealed the presence of mainly CO_2, CO, propane, and hydrogen. Oxygen was not present. The hydrocracking of canola oil on a NiMo/Al_2O_3 catalyst produced a diesel fraction mainly composed of normal paraffins from C_{15} to C_{18}. Traces of unconverted fatty acids and a very low amount of water were also detected. After evaluation of the composition and physical properties of the liquid product, they were found to be very similar to those of petro diesel, but with the advantage that the former contains a higher percentage of C_{17} and C_{18} hydrocarbons, and therefore a higher cetane number.

Keywords: *Hydrocracking, hydrotreating, vegetable oil, canola, NiMo/Al_2O_3 catalyst.*

1. Introduction

The efficient utilization of biomass for the production of transportation fuels such as biodiesel and bioethanol is becoming increasingly important. Biofuels have the potential to replace a large amount of the current gasoline and diesel used worldwide. The use of biofuels can help to slow down the consumption of petroleum while preserving the oil reserves. An important characteristic in the use of biofuels is that during their combustion there is a neutral carbon dioxide balance, which has a reduced effect on global warming. The use of transportation biofuels can also improve rural environment, generate jobs, and reduce the dependency of imported fuels.

Biodiesel as a prospective important transportation fuel is mainly produced by transesterification. A large production of biodiesel involves a large usage of water and the byproduct glycerol needs to be treated to be reused. In addition, the oxidation stability of the biodiesel is poor, and the engine may need some modifications or special maintenance when methyl esters are used. Moreover new plants are required to be built.

The above drawbacks can be overcome by using the current technology found in refinery plants in which vacuum gas oil (VGO) is hydrotreated. The basic idea is to mix the VGO stream with vegetable oil and by effect of a high pressure, high temperature and a bifunctional catalyst the triglycerides in the oil are transformed into hydrocarbons in the diesel boiling range. The metal function of the catalyst and a high hydrogen pressure contribute to the saturation of the side chains of the triglycerides. The acid function of the catalyst contributes to the cracking of the C-O bound and to the isomerization of the n-olefins formed, which are then transformed in isoparaffins. A high enough temperature is important to increase the cracking activity. However at temperatures higher than 380 °C cracking of the hydrocarbons increases and hence the yield of diesel decreases. Though more gasoline and propane can be obtained in this way, and depending of the refiner requirements, they may also be desirable products.

The hydrocarbon mixture produced from the hydrocracking of vegetable oil is commonly called green diesel, as it is indeed diesel but produced from a green plant. Green diesel is mainly composed of n-heptadecane and n-octadecane. As both of them contain a high cetane number, green diesel can also be used as a cetane additive. A green diesel with high content of isoparaffins is desirable as they have lower pour point than those of the corresponding n-paraffins. The yield of isoparaffins depends on the acid sites of the catalyst. Thus, it becomes important to study different operative conditions and different catalysts to elucidate the operative range than allows obtaining a high yield of high quality green diesel.

A number of papers[1-3] and patents[4-7] related with hydroconversion of vegetable oils have been previously published. Most efforts have focused on experiments with sulfided NiMo/γ-Al_2O_3 catalysts. This catalyst is commonly used at petroleum refineries for the hydroprocessing of middle distillates. The high hydrogenation activity and mild acidity make this catalyst appropriate for the hydroconversion of vegetable oils. More acid-supported catalysts such as Pt/H-ZSM-5 and Pt/USY zeolite, Pt/SAPO11, Pt/HZSM-22 have also been used for the same purpose[7]. The oils that have been commonly used are sunflower, rapeseed and some others like babacu oil. Huber and Corma[2] have previously shown the effects of mixing VGO with sunflower oil.

In this study, NiMo/γ-Al_2O_3 catalysts and Pt/zeolite catalyst are used for the hydrocracking of canola oil. Several reaction temperatures and hydrogen pressures and three types of catalysts have been studied to provide insights that allow for a future optimal operation.

2. Chemistry of the hydroconversion of vegetable oils

Based on their analytical results, several authors have tried to elucidate the mechanism of the hydrocracking of vegetable oils into diesel. It is generally acknowledged that the triglycerides are first saturated on their side chain, which is then followed by scission of the C-O bound, leading to the formation of diglycerides, monoglycerides, carboxylic acids and waxes[2]:

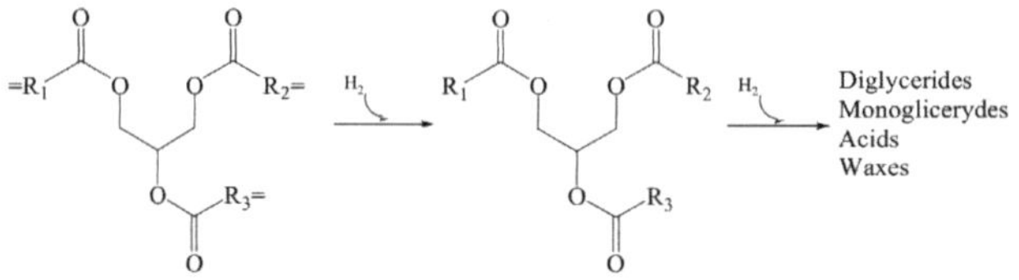

Then, these are transformed into hydrocarbons by three different paths, i.e. hydrodeoxygenation, decarbonylation and dehydration:

a. Decarboxylation:

$$C_{17}H_{35}COOH \longrightarrow C_{17}H_{36} + CO_2$$

b. Decarbonylation:

$$C_{17}H_{35}COOH + H_2 \longrightarrow C_{17}H_{36} + H_2O + CO \quad \text{(Catalytic reaction)}$$

$$C_{17}H_{35}COOH \longrightarrow C_{15}H_{31}CH{=}CH_2 + H_2O + CO \quad \text{(Thermal reaction)}$$

c. Reduction (hydrodeoxygenation):

$$C_{17}H_{35}COOH + 3H_2 \longrightarrow C_{18}H_{38} + 2H_2O$$

The predominant hydroconverted products are n-heptadecane and n-octadecane. As byproducts, carbon monoxide, carbon dioxide, and water are obtained. Due to the acid function of the catalyst, isomerization and cyclization of the olefin intermediates can occur, leading to the production of isoparaffins, naphthenes and in case there is not enough hydrogenation, aromatics can be formed. Dehydrogenation is favored at high temperature and low pressure.

3. Methods

3.1 Materials

The canola oil used in the experiments corresponds to a commercial type sold in Japan. Table 1 shows the total fatty acid composition of the oil. This was determined by derivation of the corresponding fatty acid methyl esters (FAME). A modified method AOAC 969.33 by Lee et al.[8] was used to obtain the FAME. A GC-2014 Shimadzu was used. Table 2 shows some physical properties of the oil. The CHNS composition was determined by using an elemental analyzer (CE instruments EA1110), the oxygen content was obtained in an approximated way by balance.

The density was determined at 20 °C using a density/specific gravity meter (Kyoto Electronics DA-130N). The viscosity of the oil was determined at 20 and 40 °C using a vibro viscometer (A&D Co. Lim. Japan, SV-10). The acid value of canola oil was obtained by titration with a KOH solution (0.1 M).

Table 1. Fatty acid composition of canola oil.

Fatty acid	Structure*	Wt%
Miristic	C14:0	0.00
Palmitic	C16:0	3.65
Palmitoleic	C16:1	0.18
Stearic	C18:0	1.65
Oleic	C18:1	63.72
Linoleic	C18:2	15.42
Linolenic	C18:3	14.28
Cis-11-eicosenoic	C20:1	1.11

*Cx:y, where x: number of carbon atoms; y: number of double bonds

Table 2. Physicochemical properties of canola oil

Property		
Elemental composition (%)		
	C	77.903
	H	11.689
	N	0.041
	S	0.000
	O*	10.367
Density at 20 °C , g cm^{-3}		0.89
Viscosity at 25 °C, mPa-s		63.0
Viscosity at 40 °C, mPa-s		28.1
Acid value		0.55
FFA content, wt%		0.28

*By balance

3.2. Catalysts

NiMo/γ-alumina

The catalyst precursor consisting in a mixture of NiO and MoO supported in γ-alumina corresponded to a commercial type (CDS-R25NQ) and was supplied by Catalyst and Chemicals Ind. Co. This catalyst is used in refining operations for hydrotreating of gas oil and atmospheric residue. It contains a high desulfurization activity. The activation of the catalyst was carried out by two methods: A) in situ activation with elemental sulphur: a ratio of 0.8 g S/g catalyst was used. B) Presulfidation in a tube reactor in which the catalyst was sulfided with H_2S (10 vol. %)/H_2 gas mixture at 400 °C for 8 hours and a hydrogen flow rate of 50 mL (STP)/min. The final catalyst was crushed in particles of about 0.3 mm of diameter either before method A or after method B.

Pt/HZSM5 catalyst (2 wt% Pt)

The precursor NH4-HZSM-5 was obtained by ionic exchange between Na-ZSM-5 and ammonium chloride (NH_4Cl). This precursor was dried for 24 hours at 110 °C and then calcined at 550 °C in air for three hours to form HZSM-5. The HZSM-5 particle was then impregnated with a 1 wt% $H_2PtCl_6{\cdot}6\ H_2O$ solution by incipient wetness method. The solution was added dropwise to the particle. For a catalyst containing 2% of platinum, 1.5 g of HZSM-5 powder and 7.95 g of $H_2PtCl_6{\cdot}6\ H_2O$ solution were used. About 20 mL of water were also added. The mixture then was heated at 30 °C and stirred for one hour so the platinum was absorbed into the zeolite pores. The solution was then dried at 95 °C for 24 hours. Subsequently, the particle was calcined at 400 °C in air for three hours. Finally, the catalyst sample was reduced at 350 °C, for two hours, in a reactor tube with an internal diameter of 1 cm, and a hydrogen flow rate of 50 mL (STP)/min.

Pt/H-Y zeolite catalyst (2 wt% Pt)

This catalyst corresponded to a synthetic HS-320 catalyst and was purchased from Waco Chemicals Co. (I.D. 325-27765). The SiO_2/Al_2O_3 molar ratio is 5.5.

3.3. Experimental procedure

Experiments were conducted in an 80 mL-batch reactor with an internal diameter of 20 mm, and equipped with a mechanical stirrer. The operative limits of the reactor were 30 MPa and 900 °C. For all experiments, the stirrer speed was kept constant at 350 rpm, and the temperature control was ± 2 °C.

The feed consisted in canola oil and catalyst in a ratio of 3 wt % cat/wt. oil. The reaction conditions for hydrotreating experiments were in a temperature range of 300 to 400°C, initial hydrogen pressures between 5 and 11 MPa. to reaction, the reactor was purged several times with nitrogen and then with hydrogen at room temperature. The reactor was maintained at the final temperature for 3 hours (reaction time).

After cooling down the reactor, the gas and liquid products were recovered. The liquid phase containing the catalyst was filtered. When solid product was obtained this was diluted prior to its analysis with 1,2,3,4-tetra hydro naphthalene. Both gas and liquid phases were analyzed by gas chromatography. To verify the presence in the liquid product of important compounds, such as remaining carboxylic acids, isoparaffins, cyclic paraffins and aromatics, a detailed analysis by mass spectrometry was performed. A small amount of water (ca. 4 wt%/wt oil) was also observed in the liquid phase. The produced gases were analyzed by a GC 323 (GL Sciences) equipped with a thermal conductivity detector and two columns, one being a Pora-Q capillary column (30 m, 0.53 mm I.D.) for determination of CO_2, and the other one a packed column (MS-5A) for determination of H_2,O_2, N_2 and CO. The hydrocarbon fraction C_4- was determined by a FID GC-353B (GL Sciences).

4. Results and discussion

4.1. Canola oil composition

The composition of canola oil in Table 1, reveals a high content of unsaturated carboxylic acids, i.e. oleic, linoleic and linolenic acids. Conversely, the content of saturated acids is very low. A high content of unsaturated acids implies a high consumption of hydrogen during hydrocracking process.

In Table 2 it is also possible to notice that this commercial canola oil does not contain a significant amount of free fatty acids (FFA), as it is given by the acid value. This low content of FFA is characteristic of commercial oils. The low content of FFA is desirable since the higher this value the higher the consumption of hydrogen required and therefore the higher the operative cost.

4.2. Product composition

Figure 1 shows a typical composition of the liquid product. This case corresponds to the liquid obtained at an initial pressure of 9 MPa, final temperature of 350 °C and using $NiMo/Al_2O_3$ as catalyst. With this catalyst, the hydrocracking of canola oil produces mainly normal paraffins from C_{15} to C_{20}, which are in the interval of the diesel fraction obtained from distillation of petroleum. At this relatively low temperature, a non-significant amount of gasoline (C_5-C_{12} hydrocarbons) is produced.

The green diesel mainly contains C_{17} and C_{18} normal paraffins. The content of isoparaffins (viz. Fig. 2) is very low. The main isoparaffins observed by GC-MS analysis were single-branched paraffins, e.g. 1-methyl and 2-methyl paraffins.

As the cetane number of these normal paraffins is higher than 100, the cetane number of the green diesel will be very high, and probably much higher than that of biodiesel or that of petrodiesel.

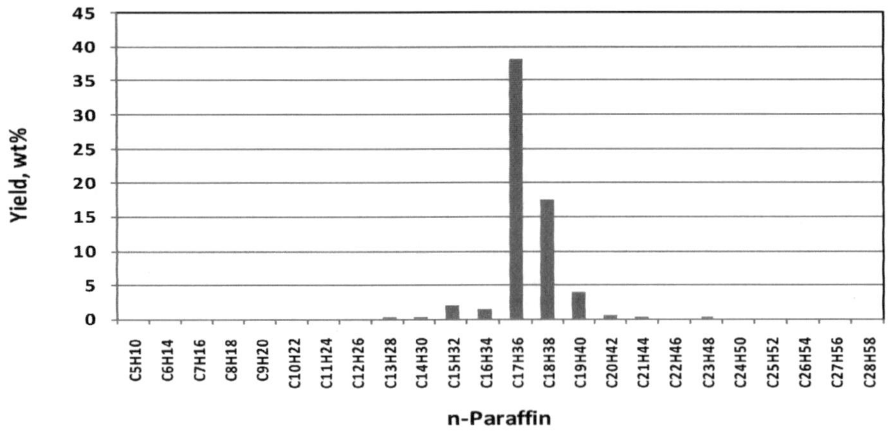

Figure 1. Yields of the C_5-C_{20} n-paraffins in the liquid fraction obtained with $NiMo/Al_2O_3$ as catalyst.

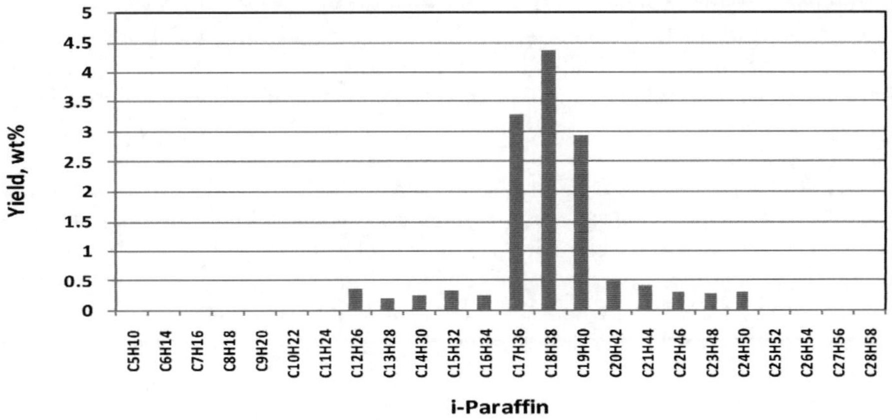

Figure 2. Yields of the C_5-C_{28} i-paraffins in the liquid fraction obtained with $NiMo/Al_2O_3$ as catalyst.

To increase the production of gasoline, temperatures higher than 350 °C can be used. Cracking of the formed paraffins is thus increased and the yield of gasoline increases as well, although the yield of diesel will be reduced.

By GC-MS analysis it was also detected the presence of cycloalkanes and aromatics. The latter ones were more predominant in the products obtained with zeolitic catalysts. This may be due to the strong acids sites in the zeolites that promote isomerization and cyclization.

Figure 3 shows the components in the gas phase (excluding the remaining hydrogen (ca. 40% mol). The yield of CO_2 predominates over CO, which indicates that decarboxylation is favored over decarbonylation. The latter reaction requires more hydrogen than the former one. As can be observed in Figure 3, propane is also produced, which in a large scale operation it will be very beneficial.

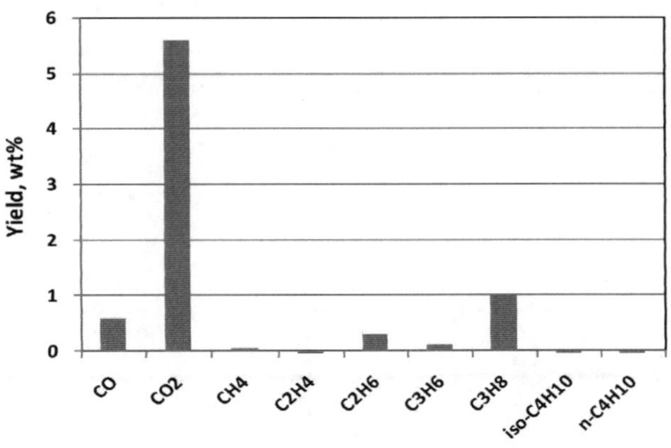

Figure 3. Yields of the components in the gas phase recovered after three-hour reaction at 350 °C with $NiMo/Al_2O_3$ as catalyst.

4.3. Effect of initial hydrogen pressure

As the hydrogen consumption is one of the important limiting economical factors in hydrotreating, the effect of the initial hydrogen pressure was studied. To evaluate this effect several experiments were performed in an interval of 5 to 11 MPa. When using $NiMo/Al_2O_3$ as catalyst, initial hydrogen pressures of 8 to 10 were found to be more appropriate. At pressures below 8 MPa the product was mostly solid and a high content of carboxylic acids were detected. They were mainly palmitic, stearic and oleic acids. With zeolitic catalysts, even higher pressures were required. In this section, only the results with $NiMo/Al_2O_3$ as catalyst are shown. In section 4.5, the results with zeolitic catalyst are described.

Figures 4, 5 and 7 show the global yields for the different fractions obtained when using $NiMo/Al_2O_3$, a reaction time of three hours at 350 °C, and different initial hydrogen pressures. The main effect of the initial pressure is on the yield of the wax fraction and on the yield of fatty acids. The global yield of green diesel is not affected significantly by the pressure. Figure 6 shows the yields of the two main components in the green diesel, i.e. n-heptadecane and n-octadecane. The higher yield of the former and the high content of CO_2 in the gas phase indicate that the main reaction path is decarboxylation.

The effect on increasing the initial hydrogen pressure is slightly more pronounced on the yield of n-octadecane than on the yield of n-heptadecane. Thus, it is posssible to argue that hydrodeoxygenation is favored at higher pressures This however produces more water as well, which may deactivate the catalyst and reduce the yield of green diesel. Figure 7 shows the global yield of carboxylic acids obtained at different initial hydrogen pressures. Even at 10 MPa, there fatty acids, mainly saturated ones, are not totally converted into hydrocarbons. To achive that, it is necessary to increase the temperature as described in the next section.

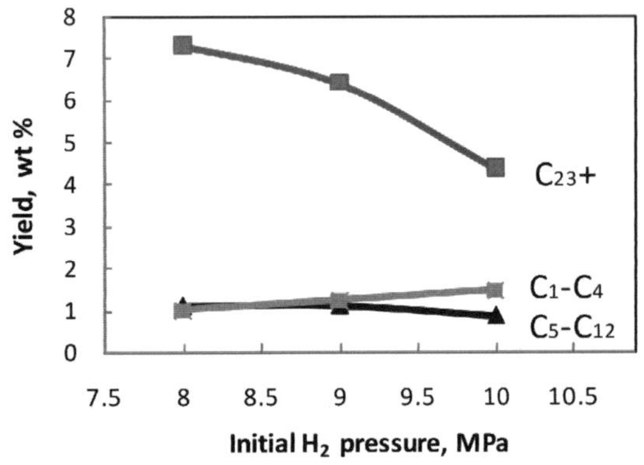

Figure 4. Influence of initial hydrogen pressure on the production of the global yields of the main fractions: gas (C_1-C_4), gasoline (C_5-C_{12}), and wax (C_{23}^{+}). T = 350 °C on $NiMo/Al_2O_3$.

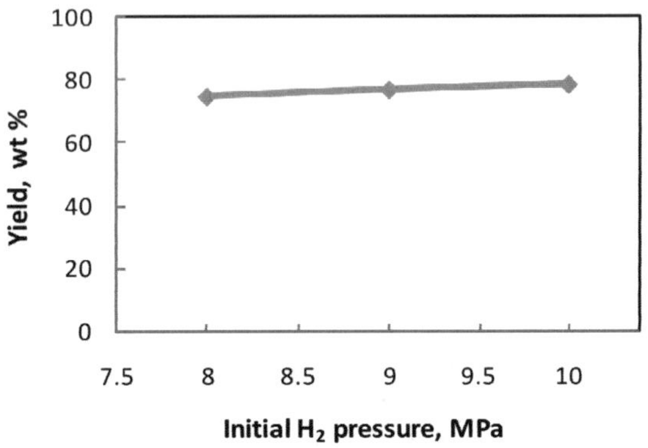

Figure 5. Influence of initial hydrogen pressure on the production of the global yields of the main fraction: green diesel (C_{12}-C_{22})

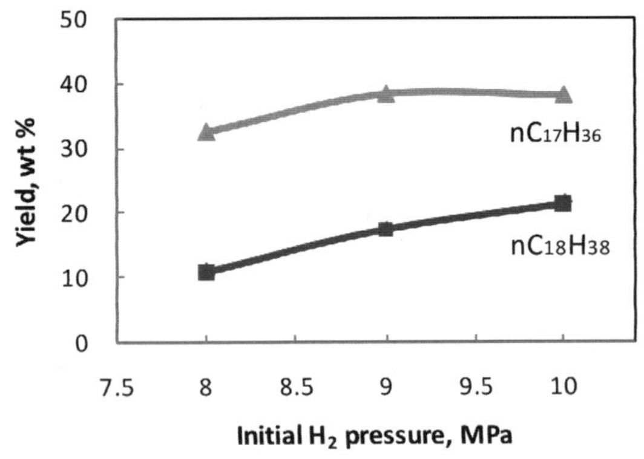

Figure 6 Influence of initial hydrogen pressure for the major products: n-heptadecane and n-octadecane. T = 350 °C on $NiMo/Al_2O_3$.

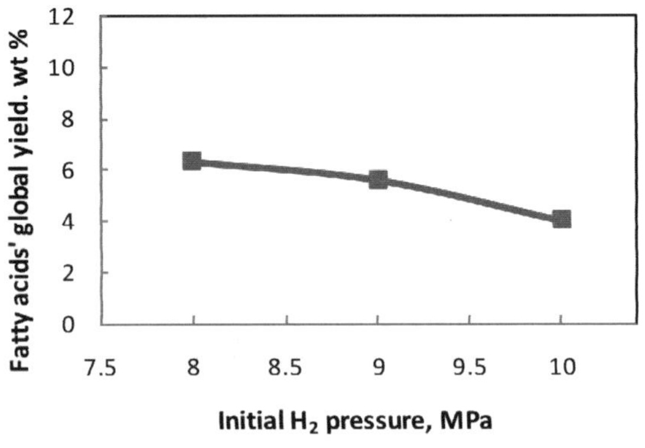

Figure 7. Influence of initial hydrogen pressure for the major products: n-heptadecane and n-octadecane. T = 350 °C on $NiMo/Al_2O_3$.

4.4. Effect of temperature

The temperature plays an important role in cracking the C=O bound. The effect of temperature on the hydrocracking of canola oil on $NiMo/Al_2O_3$ was studied by performing experiments from 300 to 400 °C , and fixing the initial hydrogen pressure at 9 MPa. The reaction time was three hours. Initial experiments from 1 to 6 hours at

350 °C were carried out to determine the optimal reaction time. Three hours were found to be appropriate. Higher reaction time did not increase the yield of the main components; conversely they were slightly reduced as cracking occurred.

Global yields of the different fractions obtained at different temperatures can be observed in Figures 8 and 9. The yields of the main products are shown in Figure 10. At temperatures below 350 °C the product was mostly solid. Only a small amount of liquid was obtained. Therefore only those results obtained at temperatures at 350 °C and above are presented here. At temperatures above 350 °C there is an increase in the degree of cracking of n-paraffins and isoparaffins in the boiling range of diesel fraction, and therefore the yield of C_5 to C_{12} (gasoline fraction) increases..

The highest yield of the C_{13}-C_{22} fraction is observed at 350 °C. The global yield of fatty acids in the product obtained at differet temperatures is also observed in Fig. 11. These results indicate that an enough high temperature (375 °C) is required to hydrocrack the carboxylic acids. The transformation of carboxylic acids can be considered therefore as the most important step for the production of diesel or gasoline hydrocarbons from the hydrocracking of vegetable oils.

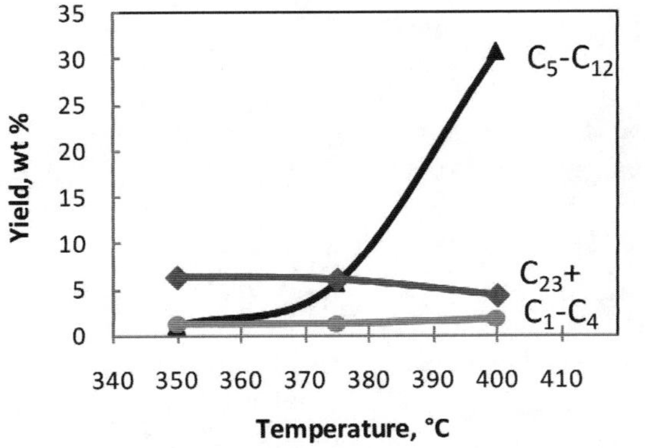

Figure 8. Influence of temperature on the production of the main fractions: gas (C_1-C_4), gasoline (C_5-C_{12}), and wax (C_{23}^{+}). Initial P_{H2} = 9 MPa. NiMo/Al_2O_3 as catalyst.

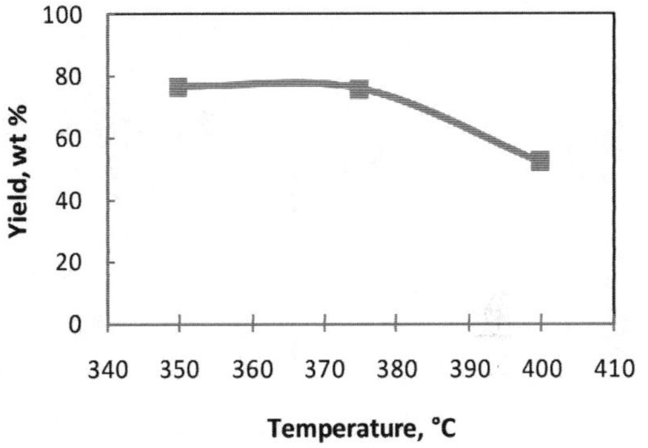

Figure 9. Influence of temperature on the production of green diesel (C_{13}-C_{22}) Initial P_{H2} = 9 MPa. NiMo/Al_2O_3 as catalyst.

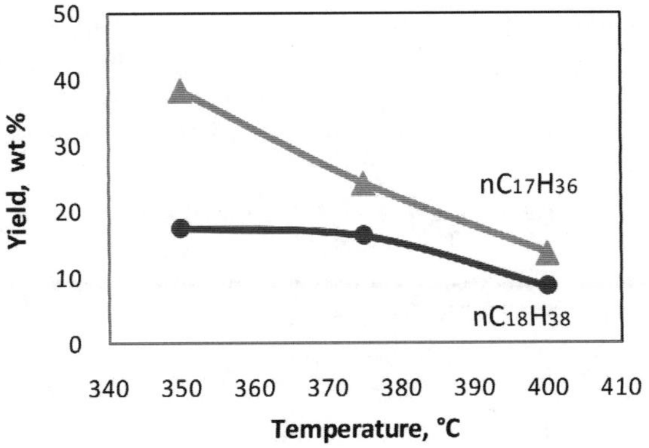

Figure 10. Influence of temperature on the production of major products. Initial P_{H2} = 9 MPa. NiMo/Al_2O_3 as catalyst.

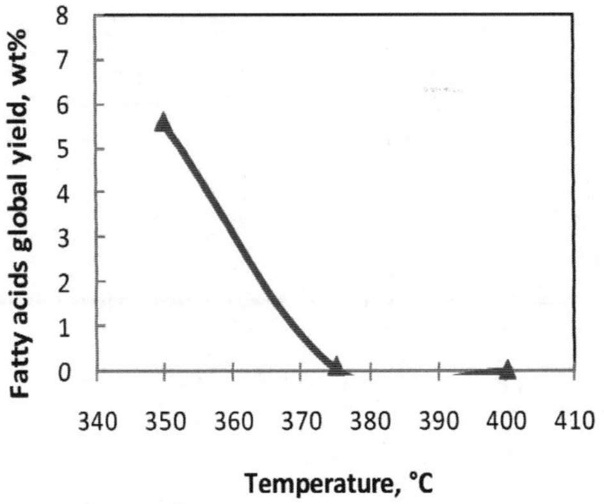

Figure 11. Influence of temperature on the global yield of fatty acids. Initial P_{H2} = 9 MPa. NiMo/Al_2O_3 as catalyst.

4.5. Zeolitic catalyst performance

The high content of C_{15} toC_{18} n-paraffins assures that the green diesel obtained with NiMo/Al_2O_3 as catalyst has a very high cetane number. However the pour point of these n-paraffins is relatively high and therefore the use of pure green diesel might not be recommendable for inter season. The best option is to transformed these n-paraffins into their corresponding isoparaffins, which have a lower pour point. As the isomerization reaction is promoted by the acids sites of the catalyst, the use of strong acid catalyst, such as zeolites can be appropriate.

Two types of Pt/zeolite catalysts were used, i.e. Pt/HZSM5 and Pt/H-Y. Figure 12 compares the yields of the major fractions obtained with both catalyst when operating with an initial hydrogen pressure of 11 MPa and 380 °C . These conditions were the ones that gave the highest yield of diesel from both catalysts. Pt/HZSM5 is observed to have more cracking activity than Pt/H-Y as it produces more gasoline and gas fractions. Pt-HY presents a lower cracking activity and the yield of diesel obtained with it is higher than that obtained with Pt-HZSM5. However, compared with the yield of diesel obtained with $NiMo/Al_2O_3$ (ca. 80%), the yield with Pt-HY is very low (ca. 35%). The ratio of isoparaffins to n-paraffins to with zeolitic catalyst is larger than that of the product obtained with $NiMo/Al_2O_3$. Which has the benefit of a lower pour point.

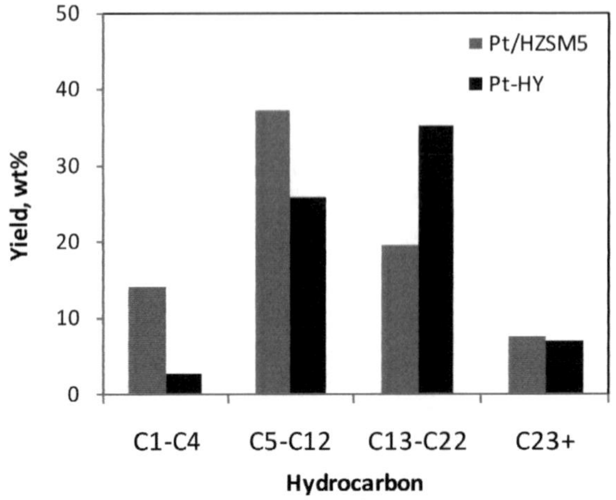

Figure 12. Global yields obtained with the operation of the two zeolitic catalyst used in the experiments. Initial P_{H2} = 11 MPa. Final T= 380 °C.

The results obtained with this study indicate that both pressure and temperature play an important role in the transformation of triglycerides and mainly carboxylic acids. An enough high pressure is needed to hydrogenate the unsaturated chains in the triglycerides and in the carboxylic acids as well as those of the olefins formed by cracking. Increasing temperature favors the cracking of the double bond C=O in the carboxylic acids, thus favoring the production of n-paraffins.

5. Conclusions

The hydrotreating of vegetable oils on bifunctional catalysts makes possible the production of liquid hydrocarbons known as "green" diesel containing mostly n-heptadecane and n-octadecane, which are mainly formed by decarbonylation and hydrodeoxygenation reactions, respectively. Green diesel can be used directly as transportation fuel or used as an excellent additive for increasing the cetane number of the petro diesel.

Due to their strong acid sites, the Pt/zeolites-supported catalysts have a high cracking activity. The hydrogenation activity is also very high and therefore the hydrogen is very fast consumed with these catalysts. In consequence, the yield of diesel is lower than that obtained when using $NiMo/Al_2O_3$ as catalyst. The process to produce diesel with the former requires higher severity, and therefore a higher cost than the latter. However, due to their strong acid sites, the use of zeolite catalyst favors the production of isoparaffins, which are desirable for the diesel to have a low pour point. It is therefore necessary to moderate the acidity of the zeolite catalysts to increase the isomerization activity while keeping a moderate cracking. Thus a large amount of isoparaffins could be obtained.

The transformation of carboxylic acids is the most important step for the production of diesel hydrocarbons from canola oil and similar oils. The quality of the diesel product depends mainly on the pressure, temperature and in the catalyst used. NiMo catalyst effectively catalyzes the hydrocracking of triglycerides and carboxylic acids lower temperature and pressure, however the moderate acidity of the alumina support does not largely contribute to the production of isoparaffins, which are also important due to their lower melting point than their corresponding n-paraffins.

Due to its large concentration of nC_{17} and nC_{18} hydrocarbons, the green diesel has a high cetane number even larger than that of the typical biodiesel composed by FAME. It can therefore be used as an excellent additive for improving the ignition properties of the petro diesel.

Acknowledments

The Japanese International Cooperation Agency (JICA) is acknowledged for the financial support to carry out the present study. COFAA of the Instituto Politecnico Nacional of Mexico is also acknowledged for providing financial support for the presentation of this work at the AIChE 2008 Annual Meeting.

References

1. Da Rocha Filho, G. N.; Brodzki, D.; Djega-Mariadassou, G. Formation of Alkanes, Alkylcycloalkanes and Alkylbenzenes During Catalytic Hydrocracking of Vegetable Oils. *Fuel* **1993,** *72*, 543-549.

2. Huber, G. W.; O'Connor, P.; Corma, A. Processing Biomass in Conventional Oil Refineries: Production of High Quality Diesel by Hydrotreating Vegetable Oils in Heavy Vaccum Oil Mixtures. *Applied Catalysis A: General* **2007**, 120-129.

3. Gusmao, J.; Brodzki, D.; Djega-Mariadassou, G.; Frety, R. Utilization of Vegetable Oils as an Alternative Source for Diesel-Type Fuel: Hydrocracking on Reduced Ni/Sio2 and Sulphided Nimo/Al2O3. *Catalysis Today* **1989**, 533-544.

4. Craig, W. K.; Soveran, D. W. Production of Hydrocarbons with a Relatively High Cetane Rating. 4992605, Feb. 12, 1991, 1991.

5. Herskowitz, M.; Landau, M.; Reizner, I.; Kaliya, M. Production of Diesel from Vegetable and Animal Oils. US 2006/0207166 A1, Sep. 21, 2006, 2006.

6. Gomes, J. R. Vegetable Oil Hydroconversion Process. US 2006/0186020 A1, Aug. 24, 2006, 2006.

7. Duarte, F. A.; Soares, W.; Ferreira, A. E.; Ximenes, L. M.; Gomes, J. R. Process to Obtain N-Paraffins from Vegetable Oil. US 2007/0260102 A1, Nov. 8, 2007, 2007.

8. Lee, D. S.; Noh, B. S.; Bae, S. Y.; Kim, K. Characterization of Fatty Acids Composition in Vegetable Oils by Gas Chromatography and Chemometrics. *Analytica Chimica Acta* **1998**, 163-175.

689e Simultaneous Transesterification and Esterification to Biodiesel Using Zinc-Based Catalyst

Shuli Yan[1], Manhoe Kim[2], John Wilson[3], Steven O. Salley[4], and K. Y. Simon Ng[4]. (1) Wayne State University, 5050 Anthony Wayne Drive, Detroit, MI 48202, (2) National Biofuel Energy Laboratory, NextEnergy, USA, 461 Burroughs Street, Detroit, MI 48202, (3) National Biofuel Energy Laboratory, NextEnergy, 461 Burroughs ST, Detroit, MI 48202, (4) Department of Chemical Engineering and Materials Science, Wayne State University, 5050 Anthony Wayne Drive, Detroit, MI 48202

Biodiesel is a mixture of fatty acid esters which can be produced from vegetable oils or animal fats with methanol. Recently, many researchers focus on using waste or unrefined oils as feedstock to decrease the production cost of biodiesel. A two-step method was reported, which firstly esterified free fatty acids (FFA) with methanol in the presence of H2SO4, then transesterified oil with methanol in the presence of NaOH. This production process is long and highly corrosive. In this study, a single-step method was developed for biodiesel production using waste or unrefined oils as feedstock based on a series of heterogeneous zinc and lanthanum mixed oxides. Oil transesterification and FFA esterification reactions simultaneously took place in one reactor. In this study, effects of catalyst structure, metal oxide molar ratio, FFA and water contents in feedstock, reaction temperature and oil concentration on the yield of biodiesel were investigated. The zinc and lanthanum catalysts were prepared by a homogeneous coprecipitation method using urea as precipitant, and characterized by XRD, XPS and EDS measurements. Results indicated that there was a strong interaction between Zn and La species which varied with the molar ratio of Zn to La. At a high ratio of Zn to La, La acted as a diluent of the matrix, promoting ZnO particle distribution, increasing the surface basic and acid sites, and enhancing activity of transesterification and esterification. At a low ratio of Zn to La, destruction of ZnO crystal structure by La species was observed and the catalytic activity was decreased. A reaction temperature window was found in 170 ~ 220 oC for biodiesel formation. At the optimal reaction conditions, 42:1 molar ratio of methanol to oil, 2.3 (wt) % catalyst and an agitation rate of 400 rpm, a high yield (95 %) of biodiesel was obtained within 3 hours using some kinds of waste or unrefined oils as feedstock. This class of zinc and lanthanum mixed oxides catalysts are very promising to lower the overall production cost of biodiesel.

689f Identification of Metals Found In Biofuel Lipids Using Inductively Coupled Plasma/mass Spectrometry

Tracy J. Benson[1], Dr. Rafael Hernandez[2], Todd French[1], William E. Holmes[3], and Ashli Brown[4]. (1) Chemical Engineering, Mississippi State University, Mississippi State, MS 39759, (2) Dave C. Swalm School of Chemical Engineering, Mississippi State University, Box 9595, Mississippi State, MS 39762, (3) Mississippi State Chemical Laboratory, Mississippi State University, Box 9595, Mississippi State University, MS 39762, (4) Biochemistry and Molecular Biology, Mississippi State University, Dorman Hall, Mississippi State, MS 39762

Recent research has identified the use of lipids as feedstocks for the production of green fuels via heterogeneous catalytic cracking. These green fuels, which are comprised of the same types of compounds found in petroleum-based fuels, could be produced from a wide variety of plant, animal, and microbial lipid sources and could replace significant amounts of petroleum fuels. The lipids are comprised mainly of free fatty acids, acylglycerols, and phospholipids and contain trace levels of metals either from growth media or from extraction processes. High phosphorus levels are expected due to the phospholipid content of extracted oils.

The metals, lithium, sodium, magnesium, phosphorus, potassium, calcium, strontium, and barium, have been identified as poisons to the heterogeneous solid catalysts used in catalytic cracking processes. A catalyst poison is defined by any agent that lowers the effectiveness of the solid substrates. This can include coke, water, and metals. Literature has shown that the above stated metals can hinder catalytic activities and even alter product distributions. Therefore, clean-up strategies must be employed to remove metals from the lipid feedstocks before catalytic cracking.

This paper will demonstrate the utility of using Inductively Coupled Plasma/Mass Spectrometry (ICP/MS) as an analytical tool for inorganic analysis of catalyst poisoning agents. In this study, lipids produced and extracted from oleaginous yeasts have been examined using ICP/MS to identify and quantitate these metals and other potentially catalyst poisoning metals. Metals analysis was performed on the bulk lipid, which included both neutral and polar lipids, separated fractions of the bulk lipid material, and the growth media. First, a full mass spectral scan was used to characterize a wide range of metals. Then, specific metals were quantitated based upon their respective dominate mass ions. An assessment has been made on the quantity of the uptake of metals from the growth media to the extracted lipids. Results from this work will be used to develop clean-up strategies for lipids before catalytic processing.

696a Sustainable Biomass Based Biorefinery

Manoj Kumar, White Biotechnology, DSM Innovation Center Inc, 45 Waterview Blvd, Parsippany, NJ 07054

DSM brings a sustainable operational history and an industry leading skill-set in the field of Industrial Biotechnology to the challenge of Biorefining. We are leveraging the longstanding competencies and knowledge derived from the biotechnology, chemistry, and operational components of DSM.

In this presentation, we will describe our vision for sustainable biorefining based upon full integration of all operational and technology elements to be developed concurrently within powerful industry partnerships.

We will provide data to illustrate that today's business model for the supply of bioproducts software into the Bioethanol/biofuels/Bio-based chemicals and materials market sector will not be sustainable for the future, whereby such products will need to be manufactured within the boundaries of the Biorefinery itself. Bioproducts manufacturing will have to be fully integrated into such a Biorefinery or it could be connected as a “on site” or “bolt-on” or ”Over the Fence” (OTF) facility. Additional key elements of integration will include, amongst others, co-development of suitable fermentation organisms that are capable to metabolize mixed sugars simultaneously along with effective pre-treatment technologies that generate hydrolysates and or hydrolysable substrates.

We will illustrate that the above focus on integration and the ability to impact all relevant process technologies will create, through compelling economics, a strong sustainable advantage for Biorefinery practitioners on a global basis.

696b Enzymatic Hydrolysis and Ethanol Production from Afex Pretreated Bagasse at High Solid Loading Using Recombinant Ethanologens

Chandraraj Krishnan, Chemical Engineering and Material Science, Michigan State University, 2527 Engineering Building, East Lansing, MI 48824, **Venkatesh Balan**, Chemical Engineering and Materials Science, Michigan State University, 2527 Engineering Building, East Lansing, MI 48824, and Bruce Dale, Michigan State University, 2527 Engineering Building, East Lansing, MI 48824.

Sugarcane bagasse is one of the potential lignocellulosic feedstocks for cellulosic ethanol production. The ethanol yield depends on the efficiency of conversion of glycans and xylans to fermentable sugars without generating byproducts that are toxic to fermentative microorganisms. Ammonia fiber expansion (AFEX) is a promising pretreatment method which has been shown to improve enzymatic hydrolysis of polysaccharides in biomass. Here, the sugarcane bagasse was pretreated by AFEX and hydrolyzed by a mixture of cellulase, xylanase and β -glucosidase enzymes. Hydrolysis of bagasse was performed with high solid loading and the enzyme ratio was optimized for maximum glucan conversion at low enzyme load. Further, the fermentation of enzymatic hydrolysate of bagasse to ethanol was performed by using recombinant ethanologenic strains of yeast and bacteria. The hydrolysate as the sole source of carbon and nitrogen supported the growth of recombinant yeast and high level of ethanol production. Addition of yeast extract and peptone to the hydrolysate had no significant effect on growth rate and ethanol production.

696c Biological Conversion of Municipal Solid Waste to Ethanol

Jian Shi, College of Engineering-Center for Environmental Research and Technology, University of California, Riverside, 1084 Columbia Ave, Riverside, CA 92507, **Bin Yang**, University of California Riverside, 1084 Columbia Avenue, Riverside, CA 92507, and Charles E. Wyman, College of Engineering-Center for Environmental Research and Technology, University of California Riverside, 1084 Columbia Avenue, Riverside, CA 92507.

Municipal solid waste (MSW) is one of the lowest cost feedstock sources for cellulosic ethanol production. Clearly, MSW-ethanol can help address waste disposal challenges, augment the diversity of the domestic energy resource base, help mitigate the impact of potential fuel supply disruption, and improve energy security. In this study, data is reported on sugar, lignin, inhibitors, and ethanol recovery profiles for various portions of MSW including final alternative daily cover (ADC), woody waste, and ADC grass pretreated in a batch reactor with dilute sulfuric acid. Particular attention is focused on evaluating the use of our lignin blocking technology to solve problems with contamination by MSW and improve yields and costs for biological production of MSW-ethanol. In addition, techno-economic models of MSW-ethanol processing were developed to project production costs and define opportunities for improvement.

696d Design of In Situ Product Recovery Devices Using Polymeric Resins

David R. Nielsen, Department of Chemical Engineering, Massachusetts Institute of Technology, 77 Massachusetts Ave., 66-425, Cambridge, MA 02139 and Kristala L. Jones Prather, Chemical Engineering, MIT, 77 Massachusetts Ave, Cambridge, MA 02139.

The application of auxiliary phases in fermentation technology has been demonstrated for many years. The roles of auxiliary phases in integrated bioreactor designs can be versatile, and include providing protection of the biocatalyst from inhibitory substrate or metabolite concentrations as well as separation of products in situ. These functions can afford enhanced productivity, simplified product recovery and purification, and lower overall energy expenditures. In the past, considerable attention has been focused on the use of immiscible organic liquids as recovery phases, resulting in the development of a full compliment of strategies and tools useful for such designs. Alternatively, the use of solid polymer resins as recovery phases has been explored, albeit to a significantly lesser extent. As such, equivalent algorithms have yet to be developed for these designs. Here we explore the potential use of polymeric resin as in situ product recovery (ISPR) devices as applied to n-butanol fermentation by Clostridium acetobutylicum ATCC 824. We provide considerable focus on the characterization of resin materials to better understand how their composition influences performance in the present study, as well as how polymer chemistry and bulk properties can be tuned to produce ideal materials for future applications. We have demonstrated that our ISPR design can dramatically improve n-butanol production while simplifying product purification protocols and drastically reducing recovery costs.

696e Quantifying Uncertainties In Biorefinery Options – Using Expert Opinions

Farminder S. Anand, Matthew Realff, and Kristiina Iisa. Chemical and Biomolecular Engineering, Georgia Tech, 311 Ferst Drive, N.W., Atlanta, GA 30332-0100

Integrated combinations of bio-chemical and thermo-chemical routes for bio-fuel production have shown great potential [1, 2]. But, the value of all these alternatives is clouded by the scores of uncertainties surrounding various unit operations [1, 2]. Moreover, the lack of physical evidence in the form of experimental data to corroborate the uncertainties further hampers the evaluation of the various biorefinery alternatives. From decision makers point of view it becomes difficult to optimally utilize the R&D resources when one is not even able to quantify the extent of the uncertainties. In light of the above, we have used the techniques used in risk and decision analysis to elicit and aggregate expert opinions in order to obtain some initial estimates of the uncertainties. Global sensitivity analysis performed on these initial estimates of the uncertainties provides the decision maker with the relative importance of various uncertainties and the ability to optimally channel the R&D resources.

1. Anand, F. S., M. Realff, et al. (2006). Evaluating Biorefinery Options in a Pulp and Paper Mill: Carbon Balances. EPE, Engineering, Pulping & Environmental Conference. Atlanta, GA, TAPPI.

2. Anand, F. S., M. Realff, et al. (2008). Integrated Bio- & Thermo- chemical route for mixed alcohol production in a pulp and paper Mill. EPE (Engineering, Pulping & Environmental Conference) & IBBC (International Bio-energy and Bio-products Conference). Portland, OR, TAPPI.

743a Small-Scale Biorefinery Process for Sustainable Local Biomass Utilization

Kazuhiro Mochidzuki, Univ. of Tokyo, 1-8 Yayoi-cho, Inage-ku, Chiba, 263-0022, Japan, Akiyoshi Sakoda, Institute of Industrial Science, University of Tokyo, 4-6-1 Komaba, Meguro-ku, Tokyo, 153-8505, Japan, Mistuhiro Takayama, Synthetic Environment Research & Planning Institute Co.,Ltd., Matsumoto, Japan, and Yasuo Igarashi, Graduate School of Agricultural and Life Sciences, Univ. of Tokyo, Tokyo, Japan.

Local Fuel is a concept that achieves a reasonable regional energy system based on local production of fuels for local consumption from locally available feedstock. As the locally available feedstock, biomass is one of the most promising resources. Because the local fuel system is operated within a limited area, long distance transportation of both biomass feedstock and the fuels produced is not needed. Therefore, the inefficient use of transportation energy can be prevented. Furthermore, the local fuel system does not require the national-wide infrastructure to distribute uniform fuels. This self-sufficient biomass energy system meets the recent strategy for the efficient utilization of regionally-distributed biomass under a concept of Biomass Town. We have proposed and developed the local fuel system focusing on some existing regions. This presentation deals with the design and demonstration of the small-scale biorefinery process that produces the local fuel from locally available biomass. The core of the process is a small-scale bioethanol production powered by a biomass boiler. The bioethanol obtained here is used in the local transportations. The feedstocks for both bioethanol and boiler are unused biomass resources and/or energy crop obtained in the region. The residues from the bioethanol process are used as organic fertilizer. The biomass boiler produces charcoal/ash as well as the energy (steam). The charcoal/ash can be used as soil additives. The application of the organic fertilizer and soil additives creates regional material circulation.

743b Hydrolysis of Microcrystalline Cellulose In Subcritical and Supercritical Water In a Continuous Flow Reactor

Sandeep Kumar, Ayhan Demirbas, and Ram B. Gupta. Chemical Engineering, Auburn University, Auburn, AL 36849-5127

For cellulosic ethanol production, efficient hydrolysis of cellulose to easily fermentable sugars is important. Focus of this study is to maximize the yield of cellulose hydrolysis in subcritical and supercritical water at practically achievable reaction times. Microcrystalline cellulose is treated with subcritical and supercritical water in temperature range 302 - 405 °C, at 27.6 MPa, and residence times of 2.5 to 8.1 seconds. Cellulose-water slurry of 2.7 wt.% (after mixing with preheated water) is rapidly heated to the reaction temperature and the reaction product is rapidly cooled in a continuous reactor. Cellulose dissolves in water partially at 302 oC and completely at 330 °C. About 65% of cellulose converts to the oligomers and monomers at 335°C and 4.8 s, and also at 354 °C at 3.5 s. Upon increase in the reaction time or temperature to supercritical region, oligomers and monomers partially degrade to glycoaldehyde dimer, D-fructose, 1, 3 dihydroxyacetone dimer, anhydroglucose, 5-HMF and furfural. The effect of temperature, pressure, and reaction time on formation of various products is studied. In addition, the effect of base catalyst K2CO3 is examined. The catalyst increases cellulose gasification in the temperature range studied (302-333 oC).

Web Page: www.eng.auburn.edu/~gupta

743c Hydrothermal Pretreament of Rubber Wood for Ethanol Production

Yukihiko Matsumura[1], Takuya Yoshida[2], Minoru Nakashima[1], Phacharakamol Petchpradab[3], and Tawatchai Charinpanitkul[3]. (1) Hiroshima University, 1-4-1 Kagamiyama, Higashi-hiroshima-shi, Hiroshima, 739-8527, Japan, (2) Hnei, University of Hawaii at Manoa, 1680 East-West Road, POST 109, Holunolu, HI 96822, (3) Center of Excellecne in Particle Technology, Chulalongkorn University, Payathai Rd., Patumwan, Bangkok, Thailand

Rubber wood is an abundant waste in Thailand, and its proper use is needed. Meanwhile, Thailand is introducing ethanol as vehicle fuel. Thus, production of ethanol from rubber wood is a promising technology in Thailand. Collaboration between Chulalongkorn University and Hiroshima University is made for hydrothermal pretreatment, cellulase treatment, to produce glucose as feedstock of ethanol fermnetaion is conducted. Reaction rate kinetics for the rubber wood hydrothermal pretreatment was studied by changing the treatment time and temperature so that optimal condition for this treatment to maximize the glucose yield was determined. The result of this reaction rate analysis is presented.

Web Page: n/a

743d Hydrothermal Biomass Chemistry: Protein-Carbohydrate Interactions

Andrew A. Peterson, Department of Chemical Engineering, Massachusetts Institute of Technology, 77 Massachusetts Ave 66-053, Cambridge, MA 02139 and Jefferson W. Tester, Chemical Engineering Department, Massachusetts Institute of Technology, 77 Massachusetts Avenue, Room 66-454, Cambridge, MA 02139.

Biofuels can be made via processing in hydrothermal media, which have the potential of providing an efficient environment for reforming biomass to fuels. Hydrothermal media refers to processing in high-temperature (200-600C), high-pressure (3-30 MPa) sub- and supercritical water. These technologies have been demonstrated to produce a range of fuels, from gaseous fuels such as hydrogen and methane, to liquid fuels such as diesel and "biocrude". While the chemical reactions of individual species, such as glucose or cellulose, are fairly well known under these conditions, interactions between species are generally uncharacterized and can lead to unexpected reactions. We have studied the chemistry of glucose and glycine as model compounds of two of the most common categories of biomass, carbohydrates and proteins, and have confirmed a strong interaction under hydrothermal conditions similar to the well-known Maillard reaction that occurs at lower temperatures and pressures. However, we report behavior in which the reaction kinetics have an unexpected dependence on the initial concentrations of the reactants.

743e A Kinetic Model for Supercritical Water Gasification of Biomass

Fernando L. P. Resende and Phillip E. Savage. Chemical Engineering Department, University of Michigan, H.H. Dow Building, 2300, Hayward St., Ann Arbor, MI 48109-2136

The use of renewable sources of energy such as biomass is an alternative that has the potential to reduce our reliance on imported oil while addressing environmental concerns. Among the technologies that convert biomass into fuels, Supercritical Water Gasification (SCWG) has been suggested to process wet feedstocks because of the ability of water to dissolve organic components of plant materials at supercritical conditions, promoting formation of gases such as H_2 and CH_4, and limiting the amount of tar and char formed as residues.

We gasified cellulose and lignin as model compounds for biomass in supercritical water, and avoided catalytic effects from the reactor walls by using quartz reactors. The experimental results were used to develop a kinetic model able to describe formation of the gas products in SCWG. The reactions are drawn from the literature. For example, the reactions in the cellulose SCWG model include:

Cellulose Hydrolysis: $(C_6H_{10}O_5)_n + n\ H_2O \Rightarrow n\ C_6H_{12}O_6$

Glucose Decomposition: $C_6H_{12}O_6 \Rightarrow C_xH_yO_z$

Steam-reforming I: $C_xH_yO_z + (x-z)\ H_2O \Rightarrow x\ CO + (x - z + y/2)\ H_2$

Steam Reforming II: $C_xH_yO_z + (2x-z)\ H_2O \Rightarrow x\ CO_2 + (2x- z+ y/2)\ H_2$

Char formation through intermediates: $C_xH_yO_z \Rightarrow w\ C + C_{x-w}H_yO_z$

Water-Gas Shift: $CO + H_2O \Leftrightarrow CO_2 + H_2$

Methanation: $CO + 3\ H_2 \Leftrightarrow CH_4 + H_2O$

Experimental results at different temperatures, times, biomass loading and water densities were used to determine the best-fit kinetics parameters.

The kinetic parameters obtained in this model were used to extrapolate experimental results and predict which conditions optimize H_2 yield and energetic content of the product gas. This type of model can be used to simulate the performance of a SCWG plant and determine its thermal efficiency, in order to establish a comparison with current technologies for biomass processing.

743f Flash Carbonization of Sewage Sludge: Assessment of Sewage Sludge Charcoal Land Application

Takuya Yoshida and **Michael J. Antal Jr.**. Hawaii Natural Energy Institute, University of Hawaii at Manoa, 1680 East-West Road, POST 109, Holunolu, HI 96822

Large amounts of sewage sludge are produced in treatment process of wastewater worldwide, causing many problems in disposal of sewage sludge. Because sewage sludge contains nitrogen, phosphorus and inorganic materials, some of them are used as fertilizer and some are just applied to land (i.e. landfill). While the toxic material and heavy metals in the sewage sludge can cause problems when it is applied to land. There are many advantages of carbonization of sewage sludge, because carbonization can reduce the volume of sewage sludge, eliminate odor, bacteria and organic toxic substance in it. However ash and heavy metal cannot be eliminated during carbonization, which could lead serious problem.

Flash Carbonization process, which can carbonize biomass and organic material in short period, is one of options to carbonize sewage sludge. We have successfully carbonized sewage sludge in 10-15 min by means of our Flash Carbonization lab-scale reactor. The ash and heavy metal content in the sewage sludge charcoals were analyzed and their behavior during Flash Carbonization process will be discussed in the presentation. Moreover applicability assessment of sewage sludge charcoal land application will be also presented.

743g Effects of Volatile-Char Interaction on the Char Reactivity In the Biomass Gasification

Tomoko Wada, Chihiro Fushimi, and Atsushi Tsutsumi. Institute of Industrial Science, The University of Tokyo, 4-6-1 Komaba, Meguro-ku, Tokyo, 153-8505, Japan

Biomass is becoming increasingly important as a renewable source which can reduce the CO2 emission. Steam gasification is a promising technology for hydrogen production with high energy conversion efficiency. In the initial stage of biomass steam gasification, pyrolysis takes place and volatiles (tar and gases) and solid residue (char) are produced. Subsequently, steam reacts with the residual char and tar, producing gases. It is necessary to realize high conversion of the char and to accelerate the rate of gasification for achieving high overall efficiency of biomass steam gasification. Furthermore, large amount of tar is evolved in biomass gasification. The presence of tar can cause blockages and corrosion of pipes and also reduce overall thermal efficiency of the process. It is, therefore imperative to facilitate steam reforming of tar.

In coal gasification, it has been reported that interaction between the volatiles and char inhibits steam gasification of char. Similarly, it is possible that the inhibition occurs in biomass gasification. For proper design and operation of biomass gasification process, it is necessary to clarify the influence of the volatile-char interaction. However, so far there is few research that quantitatively investigates the interaction in biomass gasification.

In this research, we have developed a two-stage reactor which consists of a drop tube reactor and a thermobalance fixed-bed reactor to investigate the influence of the volatiles-char interaction on gasification rate of char derived from biomass. Initially, in the lower thermobalance fixed-bed reactor char produced by pyrolysis in advance was put into a ceramic basket. Then, after the reactor was heated and biomass sample was dropped into the upper drop tube reactor, pyrolysis started to produce char which was trapped by a filter at the middle of two reactors. Subsequently, the volatiles produced in the upper reactor was fed to the lower thermobalance fixed-bed reactor and contacted with char stocked in advance. After that, char was gasified by feeding steam and the gasification rate of char was measured. We also measured the evolution rates of gaseous products from the lower reactor using a mass spectrometer and a micro chromatograph to examine whether the decomposition of tar is facilitated at the surfaces of char.

757a Tar Reforming Catalyst Development and Scale up for Use In a Biomass Gasification Pilot Plant

Kimberly Magrini-Bair and **Whitney Jablonski**. National Bioenergy Center, National Renewable Energy Laboratory, 1617 Cole Blvd, MS 3322, Golden, CO 80401

At the National Renewable Energy Laboratory (NREL) an indirect gasification pilot plant is used to study biomass conversion to syngas. Tars contained in the raw syngas are subsequently removed via steam reforming in a fluidized bed, full stream reformer (FSR). Fluidizable catalysts developed for use in the FSR are first screened for reforming performance with a fixed-bed, bench scale micro-activity test system (MATS). The primary goal of this study was to screen catalysts of varied composition in the MATS reactor and identify the best performing catalysts for scale up and use in the pilot plant FSR for syngas cleanup. The second goal of this study was to determine the efficacy of anti-coking additives such as magnesium oxide (MgO), potassium oxide (K2O), and tin (Sn) on catalyst activity. The initial experimental regime consisted of ethylene steam reforming with a steam to carbon ratio of ~4.9 and a gas hourly space velocity (GHSV) of ~68,000 in a 0.5 inch diameter quartz tube reactor. After initial screening, the catalysts that performed best were next evaluated at double the space velocity in order to identify more subtle differences in reforming activity. Two NREL catalysts were chosen for scale up and testing in the FSR. Those catalysts were further validated in the MATS system using a model syngas feed with 20 ppm sulfur, a composition that is representative of raw syngas produced in the gasifier. Both bench and pilot scale reforming results obtained with these catalysts will be discussed.

Naphthalene Steam Reforming over Ni/$Ce_{0.75}Zr_{0.25-x}Mn_xO_2$ (x = 0, 0.10, 0.20 and 0.25) Mixed Oxide Catalysts

Vissanu Meeyoo[1,*], *Asawin Bampenrat*[2], *Boonyarach Kitiyanan*[2], *Pramoch Rangsunvigit*[2], *Thirasak Rirksomboon*[2]

[1] *Centre for Advanced Materials and Environmental Research, Mahanakorn University of Technology, Bangkok, Thailand*

[2] *The Petroleum and Petrochemical College, Chulalongkorn University, Bangkok, Thailand*

Abstract

In this study, the catalytic activity of Ni supported on $Ce_{0.75}Zr_{0.25-x}Mn_xO_2$ (x = 0, 0.10, 0.20 and 0.25) catalysts prepared by urea hydrolysis and incipient wetness impregnation is investigated via steam reforming reaction of naphthalene as a model biomass tar compound. Experiments were carried out in a fixed bed reactor at 700 °C and atmospheric pressure using a molar ratio of steam/carbon (S/C) equal to 2. The catalysts also were characterized by various techniques, such as surface area, temperature-programmed reduction (TPR), X-ray diffraction (XRD) and temperature-programmed oxidation (TPO). In comparison with Ni/α-Al_2O_3, the catalytic performance data showed that nickel supported on mixed oxide catalysts exhibit high activities and stabilities for the steam reforming of naphthalene and high resistance toward carbon deposition. Moreover, the promotion effect of manganese doping markedly suppressed carbon deposited on the Ni/$Ce_{0.75}Zr_{0.25-x}Mn_xO_2$ catalysts. It can be suggested that the incorporation of manganese ions into ceria lattice improved the oxygen storage capacity as well as the oxygen mobility on the surface of mixed oxides, facilitating gasification of surface carbon.

Introduction

Tar formation is a major problem in biomass gasification process [1]. Condensation or formation of tar aerosols cause downstream problems in the process equipments. Moreover, this tar is also carcinogenic and possibly mutagenic substance [2]. Recently, the $Ce_{0.75}Zr_{0.25}O_2$ mixed oxide solid solutions shows a high catalytic activity, particularly for the oxidation of hydrocarbons and the removal of post-combustion pollutants [3-5]. One of the important routes to improve the catalytic performances of ceria-based mixed oxides is by doping and modification with other elements. Particular attention is being paid to manganese oxides which present good oxidation ability due to several oxidation states of manganese species [6-9]. Hence, in the present work, we aimed to study the effect of manganese doping on the structural and redox properties of mixed oxides $Ce_{0.75}Zr_{0.25-x}Mn_xO_2$ (x = 0, 0.10, 0.20 and 0.25), and the catalytic activity of naphthalene (tar model) steam reforming over nickel supported on these mixed oxide.

Results and discussion

* To whom the correspondence should be addressed:
Mahanakorn University of Technology, Bangkok, Thailand 10530
Tel.:+66(0)-2988-4043; fax: +66(0)-2988-4039.
Email address: vissanu@mut.ac.th (V. Meeyoo)

From XRD spectra (Fig. 1), Ni/$Ce_{0.75}Zr_{0.25-x}Mn_xO_2$ catalysts exhibited major peaks at about 29, 33, 47, and 58° (2θ) indicating the cubic fluorite structure of CeO_2. Small peaks characteristic of NiO were observed at about 37°, 43° and 62° (2θ). No peaks of zirconia and manganese oxides are observed in the XRD patterns indicating that all zirconium cations and manganese cations dissolved in the ceria lattice. These results are similar to what reported elsewhere [9, 10]. Since the ionic radius of Zr^{4+} (8.4 nm), Mn^{4+} (5.6 nm) and Mn^{3+}(6.2 nm) are smaller than that of Ce^{4+}(9.7 nm), a dissolution within CeO_2 lattice is then possible at low dopant concentrations of both Zr and Mn. However, it was evident that at high dopant of Mn (Mn/(Mn+Ce) > 0.75), the presence of Mn_2O_3 or Mn_3O_4 was observed [7, 9, 11, 12].

The TPR results of Ni supported on $Ce_{0.75}Zr_{0.25-x}Mn_xO_2$ mixed oxide and α-Al_2O_3 catalysts are shown in Fig. 2. The reduction profiles of Ni/$Ce_{0.75}Zr_{0.25-x}Mn_xO_2$ are characterized by two peaks. The low-temperature peak at temperature of 230–350 °C which is the reduction of free NiO particles [13, 14] and another peak at higher temperature of 350–550 °C is associated with the reduction of NiO to Ni^0. It was noticed that the second reduction temperature peak of Ni/$Ce_{0.75}Mn_{0.25}O_2$ catalyst is slightly lower than that of Ni/$Ce_{0.75}Zr_{0.25-x}Mn_xO_2$ catalyst indicating a less interaction between NiO and $Ce_{0.75}Mn_{0.25}O_2$. For Ni/α-Al_2O_3 catalysts, only one broad peak was observed at 510 °C, which is attributed to agglomerated NiO.

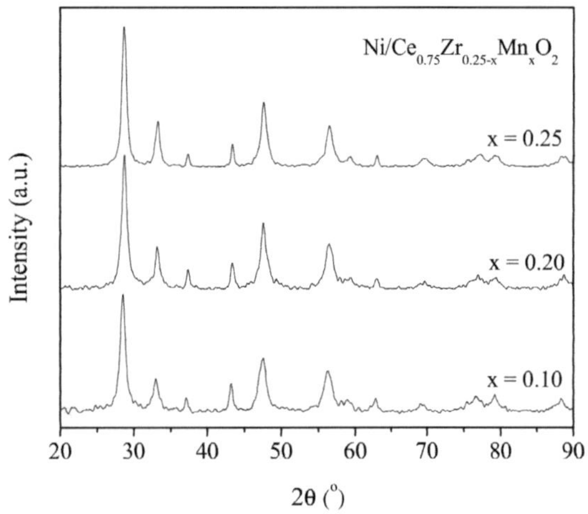

Fig. 1. X-ray diffraction patterns for Ni/$Ce_{0.75}Zr_{0.25-x}Mn_xO_2$ mixed oxide catalysts.

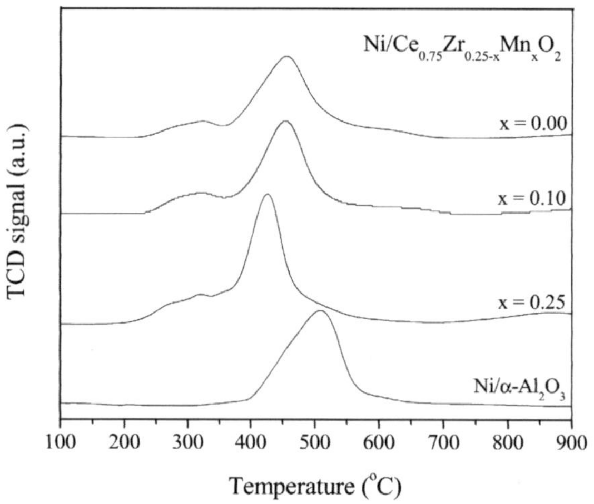

Fig. 2. H_2-TPR profiles of the Ni/$Ce_{0.75}Zr_{0.25-x}Mn_xO_2$ mixed oxide and Ni/α-Al_2O_3 catalysts.

The BET surface areas of Ni supported on $Ce_{0.75}Zr_{0.25-x}Mn_xO_2$ mixed oxide catalysts are in the range of 62–77 $m^2\,g^{-1}$ and the average pore diameters of these catalysts are 6.3–7.9 nm. It should be noted that BET surface area of the samples slightly decreases with increasing Mn content, except Ni/$Ce_{0.75}Mn_{0.25}O_2$. This is related to the formation of the large MnO_x particles whose surface area is much smaller [12, 15].

The catalytic activities and stabilities of Ni supported on $Ce_{0.75}Zr_{0.25-x}Mn_xO_2$ mixed oxide catalysts for steam reforming of naphthalene were studied and compared with those of Ni/α-Al_2O_3 catalyst. It was found that naphthalene conversions by Ni/$Ce_{0.75}Zr_{0.25-x}Mn_xO_2$ (x = 0, 0.10, 0.20, 0.25) mixed oxide catalysts are completed for at least 6 hrs time on streams, while the rapid deactivation was observed for Ni/α-Al_2O_3 catalyst after 2 hrs. This indicated that nickel supported on mixed oxide catalysts are active and more stable for steam reforming of naphthalene. This might be due to the reducibility and oxidation ability of mixed oxide supports; promoting the oxidation of carbon precursors.

Table 1 summarizes the products selectivities, hydrogen yield and H_2/CO molar ratio during naphthalene steam reforming. It was found that the steam reforming products are mainly CO, CO_2 and H_2 for $Ni/Ce_{0.75}Zr_{0.25-x}Mn_xO_2$ (x = 0.10, 0.20 and 0.25) and the H_2/CO ratio is about 2.40 ± 0.12. On the other hand, the main products of naphthalene steam reforming over $Ni/Ce_{0.75}Zr_{0.25}O_2$ are CO, CO_2, H_2 and a trace of CH_4 with the H_2/CO ratio is about 1.84 ± 0.05, indicating carbon formation. In addition, an increase in CO_2 selectivities by introducing of Mn to CeO_2-ZrO_2 mixed oxide was also observed suggesting that the water gas shift (WGS) reaction is possibly taken place over $Ni/Ce_{0.75}Zr_{0.25-x}Mn_xO_2$ (x = 0.10, 0.20 and 0.25).

Table 1. Naphthalene steam reforming over $Ni/Ce_{0.75}Zr_{0.25-x}Mn_xO_2$ mixed oxide catalysts reaction temperature = 700 °C; time on stream = 6 hrs; S/C = 2.0)

Catalyst	S_CO (%)	S_CO_2 (%)	S_CH_4 (%)	Y_H_2 (%)	H_2/CO
$Ni/Ce_{0.75}Zr_{0.25}O_2$	97.60±0.88	2.25±0.10	0.15±0.05	72.89±1.75	1.84±0.05
$Ni/Ce_{0.75}Zr_{0.15}Mn_{0.10}O_2$	79.97±1.42	20.02±0.57	0	75.36±1.77	2.28±0.09
$Ni/Ce_{0.75}Zr_{0.10}Mn_{0.15}O_2$	74.27±1.53	25.73±0.93	0	76.33±1.17	2.37±0.12
$Ni/Ce_{0.75}Mn_{0.25}O_2$	75.54±1.21	24.46±0.92	0	76.27±1.13	2.56±0.07

The carbon formation over the spent catalysts was investigated. The TPO spectra suggested two types of carbon of the same nature for Ni/α-Al_2O_3 and $Ni/Ce_{0.75}Zr_{0.25-x}Mn_xO_2$ catalysts. The first one oxidized at lower temperature ~325 °C which can be attributed to carbidic carbon and the second one oxidized at ~680 °C corresponding to filamentous carbon [16, 17]. It was found that the amounts of carbon formed on Ni/α-Al_2O_3 and $Ni/Ce_{0.75}Zr_{0.25}O_2$ catalysts are about 63.4 and 26.7 mg g_{cat}^{-1}, respectively. While $Ni/Ce_{0.75}Zr_{0.25-x}Mn_xO_2$ (x = 0.10, 0.20, 0.25) catalysts yielded much lower amounts of carbon formed (1.1 ± 0.3 mg g_{cat}^{-1}) than Ni/α-Al_2O_3 and $Ni/Ce_{0.75}Zr_{0.25}O_2$ catalysts. It can be pointed out that the total amounts of carbon decreased when manganese is added into the ceria-zirconia system.

Conclusions

In conclusion, $Ni/Ce_{0.75}Zr_{0.25-x}Mn_xO_2$ catalysts were found to be active for naphthalene steam reforming. It was found that the catalysts achieved the 100% conversion for at least 6 hrs during the course of the experiments with no sign of deactivation, unlike α-Al_2O_3, which showed the deactivation after 2 hrs times on stream. Carbon formation was found to decrease with an increase in Mn content as well as the absence of filamentous carbon. It is believed that an improvement in redox properties with the presence of Mn suppresses carbon formation.

Acknowledgements

This work was supported by the Thailand Research Fund (under Waste-to-Energy project and Royal Golden Jubilee Ph.D. Program: Grant 0170/46), and the Research Unit for Petrochemical and Environmental Catalysts, Ratchadapisek Somphot Endowment Fund, and the National Center of Excellence for Petroleum, Petrochemicals and Advanced Materials, Chulalongkorn University.

References

1. D. Sutton, B. Kelleher, J.R.H. Ross, Fuel Process. Technol. 73 (2001) 155.
2. Z. Abu El-Rub, E.A. Bramer, G. Brem, Ind. Eng. Chem. Res. 43 (2004) 6911.
3. M. Thammachart, V. Meeyoo, T. Rirksomboon, S. Osuwan, Catal. Today 68 (2001) 53.
4. S. Pengpanich, V. Meeyoo, T. Rirksomboon, K. Bunyakiat, Appl. Catal. A: Gen. 234 (2002) 221.
5. S. Pengpanich, V. Meeyoo, T. Rirksomboon, J. Schwank, Appl. Catal. A: Gen. 302 (2006) 133.
6. S. Imamura, M. Shono, N. Okamoto, A. Hamada, S. Ishida, Appl. Catal. A: Gen. 142 (1996) 279.
7. X. Wu, Q. Liang, D. Weng, J. Rare Earths 24 (2006) 549.
8. X. Wang, Q. Kang, D. Li, Catal. Comm. 9 (2008) 2158.
9. M. Machida, M. Uto, D. Kurogi, T. Kijima, Chem. Mater. 12 (2000) 3158.
10. G. Qi, R.T. Yang, R. Chang, Appl. Catal. B: Environ. 51 (2004) 93.
11. G. Qi, R.T. Yang, J. Phys. Chem. B 108 (2004) 15738.
12. G. Blanco, M.A. Cauqui. J.J. Delgado, A. Galtayries, J.A. Pérez-Omil, J.M. Rodríguez Izquierdo, Surf. Interf. Anal. 36 (2004) 752.
13. H. Roh, K. Jun, W. Dong, J. Chang, S. Park, Y. Joe, J. Mol. Catal. A 181 (2002) 137.
14. S. Pengpanich, V. Meeyoo, T. Rirksomboon, K. Bunyakiat, Catal. Today 93-95 (2004) 95.
15. X. Wu, Q. Liang, D. Weng, J. Fan, R. Ran, Catal. Today 126 (2007) 430.
16. J.C. Vargas, S. Lib, A.-C. Roger, A. Kiennemann, Catal. Today 107-108 (2005) 417.
17. D. Swierczynski, S. Libs, C. Courson, A. Kiennemann, Appl. Catal. B: Environ. 74 (2007) 211.

757c Steam Gasification Reactivity of Char from Rapid Pyrolysis of Bio-Oil/ char Slurry

Masakazu Sakaguchi, Naoko Ellis, and A. Paul Watkinson. Department of Chemical and Biological Engineering, University of British Columbia, 2360 East Mall, Vancouver, BC V6T 1Z3, Canada

Oil/char mixtures from rapid pyrolysis of wood-waste are an attractive feedstock for gasification because of their relatively high energy density, and fluid-like characteristics. Steam gasification of bio-oil/char slurry consisting of 80 wt.% of bio-oil, and 20 wt.% of char from fast pyrolysis of wood is being investigated. The char is crushed and sieved to a size below 38 μ m, and added to the bio-oil. The resulting slurry is heated to 900°C in nitrogen at nominal heating rates of 10 to 10000 K/s in a CDS Instruments Pyroprobe, in order to simulate rapid heating in a steam gasification reactor. The resulting chars from the slurry and the original char are gasified at 900°C in a thermogravimetric analyzer (TGA) with a mixed gas of 10 vol.% steam and 90 vol.% nitrogen at a total pressure of 100 kPa, and the gasification reactivity (dX/dt), where X is the fractional conversion, is determined. There are no significant differences in the steam gasification reactivities of chars originating from the slurries at the different heating rates tested. Under 10% steam at 900°C reactivity values of $0.9\text{-}1.1\times10^{-3}$ s^{-1}, $0.8\text{-}1.0\times10^{-3}$ s^{-1}, $0.7\text{-}0.9\times10^{-3}$ s^{-1} and $0.5\text{-}0.7\times10^{-3}$ s^{-1} are found at fractional conversions of 0.2, 0.4, 0.6 and 0.8, respectively. For the original char, the gasification reactivity was 1.0×10^{-3} s^{-1} at conversions of 0.2-0.6, and 0.9×10^{-3} s^{-1} at a fractional conversion of 0.8. Thus the gasification reactivities of chars made from bio-oil/char slurry are close to that of the original char, except for the final stage of the reaction at which char from the slurry is less reactive. Gasification temperature and steam concentration effects on the char gasification reactivity are also presented. A laboratory scale apparatus has been constructed for continuous steam gasification of bio-oil/char slurry.

757d Hydrogenations of Biomass-Related Substrates In Supercritical Media

Jackson W. Ford, Raghunath V. Chaudhari, and Bala Subramaniam. Center for Environmentally Beneficial Catalysis and Department of Chemical & Petroleum Engineering, University of Kansas, Lawrence, KS 66045

Biomass conversion to fuels and chemicals requires new approaches to evolve efficient and economically viable processes. Hydrogenations of biomass-related substrates can convert highly-oxygenated biomass components to feedstocks which more closely resemble petrochemicals. The high hydrogen pressures needed for such reactions adversely affect the process economics, but the use of supercritical solvents could overcome this barrier by enhancing the mutual solubilities of reactants as well as the transport of materials and energy in the reaction system. We have used a fixed bed reactor to evaluate supported transition metal catalysts and supercritical solvent media for intensified hydrogenations of biomass constituents at mild process conditions. We have used a building block approach, beginning with small-molecule biomass derivatives, to understand better the reaction mechanisms involved in biomass hydrogenations. Such an understanding is essential for developing novel methods to produce and upgrade biomass-derived feedstocks for fuels and chemicals production.

757e Fast Pyrolysis of Guayule Residues

Charles A. Mullen[1], Akwasi A. Boateng[1], Neil Goldberg[1], Kevin B. Hicks[1], Colleen McMahan[2], and Maureen Whalen[2]. (1) Eastern Regional Research Center, Agricultural Research Service, 600 Mermaid Lane, Wyndmoor, PA 19038, (2) Western Regional Research Center, Agricultural Research Service, 800 Buchanan Street, Albany, CA 94710

Guayule is a perennial shrub grown in the southwestern United States that produces hypoallergenic natural latex. However, only 10% of the plant material is used for latex production, the remaining material (bagasse) is waste biomass material that can be used for renewable fuel production. Fast pyrolysis can convert lignocellulosic biomass into pyrolysis liquids (bio-oils) that can be used as boiler fuel or can be upgraded to liquid transportation fuels. The fast pyrolysis of both guayule bagasse and the whole shrub material was studied on both an analytical and at bench scale for production of bio-oil in a fluidized bed reactor. Over a sand medium, bio-oil was produced in the 60% yield range. Bio-oils from guayule pyrolysis had energy content higher than that of typical fast pyrolysis oil, about 30 MJ/kg, 75% the value of heavy fuel oil. Results of pyrolysis over various catalysts will also be presented, along with detailed analysis of the composition and properties of the bio-oil and coproducts produced.

757f Production of Stable and up-Gradable Pyrolysis Liquid Transportation Fuel Intermediates from Lignin and Related Biomass

Akwasi A. Boateng, Charles A Mullen, Neil Goldberg, and Kevin B. Hicks. Usda, Eastern Regional Research Center, Agricultural Research Service, 600 Mermaid Lane, Wyndmoor, PA 19038

Fast pyrolysis is a liquefaction process that shows promise for converting lignocellulosic biomass to liquid fuel intermediates that can be eventually converted to liquid transportation fuels. However, raw pyrolysis oil can contain between 10 and 30 wt% water and hundreds of various oxygenated organic and reactive compounds that lead to significant challenges such as poor volatility, high viscosity, coking and corrosiveness. These attributes, in turn, limit its use as a fuel in the current transportation market. Work at ARS is aimed at combining biomass selection and pyrolysis pathway modifications to alter the composition and properties of pyrolysis oil so as to decrease some of the adverse fuel properties, increase its stability or make it more favorable for upgrading within the current refinery infrastructure. In this presentation, the quality of pyrolysis liquids produced by catalytic conversion of a 50% dry-solid extracted lignin is compared with that from a regular lignocellulosic biomass.

AUTHOR INDEX

AUTHOR INDEX

AUTHOR INDEX

AUTHOR INDEX

AUTHOR INDEX